BASICS OF MORPHOLOGY

BASICS OF MORPHOLOGY

Ramesh Umrani

ANMOL PUBLICATIONS PVT. LTD.

NEW DELHI - 110 002 (INDIA)

ANMOL PUBLICATIONS PVT. LTD.
H.O.: 4374/4B, Ansari Road, Daryaganj,
New Delhi-110 002 (India)
Ph.: 23278000, 23261597
B.O.: No. 1015, Ist Main Road, BSK IIIrd Stage
IIIrd Phase, IIIrd Block,
Bangalore - 560 085 (India)
Visit us at: www.anmolpublications.com

Basics of Morphology

First Edition, 2009
ISBN 978-81-261-3772-5

PRINTED IN INDIA

Printed at Balaji Offset Press, Delhi.

Contents

Preface

"The present title "Basics of Morphology" has been written mainly to meet the need of undergraduate and post graduate students who require a knowledge of a wide field in morphology in an interesting an understandable manner. The book presents a broad survey of the trees and forests. A special feature of the book is that the text is copiously illustrated with superably accurate, beautiful and well labelled diagrams. The subject matter has been dealt in a simple and lucid style and also included maximum of recent information available up-to-date.

This textbook represents a group effort, rather than product of any single individual. Foremost on the list stand the reviewers whose advice, comments and collective wisdom helped to shape this text into a final form. Their interest in the subject, their concern for the accuracy and method of presentation, and their experience with students of widely varying abilities and background made the review process an educational experience. To these individuals, who carefully recorded their comments, opinions, and sources. The author would like to express her sincere thanks."

Author

Chapter 1

Cell Structure

Cells are the smallest structures capable of basic life processes, such as taking in nutrients, expelling waste, and reproducing. All living things are composed of cells. Some microscopic organisms, such as bacteria and protozoa, are unicellular, meaning they consist of a single cell. Plants, animals, and fungi are multicellular; that is, they are composed of a great many cells working in concert.

But whether it makes up an entire bacterium or is just one of trillions in a human being, the cell is a marvel of design and efficiency. Cells carry out thousands of biochemical reactions each minute and reproduce new cells that perpetuate life.

Cells vary considerably in size. The smallest cell, a type of bacterium known as a mycoplasma, measures 0.0001 mm (0.000004 in) in diameter; 10,000 mycoplasmas in a row are only as wide as the diameter of a human hair. Among the largest cells are the nerve cells that run down a giraffe's neck; these cells can exceed 3 m (9.7 ft) in length. Human cells also display a variety of sizes, from small red blood cells that measure 0.00076 mm (0.00003 in) to liver cells that may be ten times larger. About 10,000 average-sized human cells can fit on the head of a pin.

Along with their differences in size, cells present an array of shapes. Some, such as the bacterium *Escherichia coli*, resemble rods. The paramecium, a type of protozoan, is slipper shaped; and the amoeba, another protozoan, has an irregular form that changes shape as it moves around. Plant cells typically resemble boxes or cubes. In humans, the outermost layers of skin cells are flat, while muscle cells are long and thin. Some nerve cells, with their elongated, tentacle-like extensions, suggest an octopus.

In multicellular organisms, shape is typically tailored to the cell's job. Flat skin cells pack tightly into a layer that protects the underlying tissues from invasion by bacteria. Long, thin muscle cells contract readily to move bones. The numerous extensions from a nerve cell enable it to connect to several other nerve cells in order to send and receive messages rapidly and efficiently.

By itself, each cell is a model of independence and self-containment. Like some miniature, walled city in perpetual rush hour, the cell constantly bustles with traffic, shuttling essential molecules from place to place to carry out the business of living. Despite their individuality, however, cells also display a remarkable ability to join, communicate, and coordinate with other cells. The human body, consists of an estimated 20 to 30 trillion cells. Dozens of different kinds of cells are organized into specialized groups called tissues. Tendons and bones, are composed of connective tissue, whereas skin and mucous membranes are built from epithelial tissue. Different tissue types are assembled into organs, which are structures specialized to perform particular functions. Examples of organs include the heart, stomach, and brain. Organs, in turn, are organized into systems such as the circulatory, digestive, or nervous systems. All together, these assembled organ systems form the human body.

The components of cells are molecules, nonliving structures formed by the union of atoms. Small molecules serve as building blocks for larger molecules. Proteins, nucleic acids, carbohydrates, and lipids, which include fats and oils, are the four major molecules that underlie cell structure and also participate in cell functions. A tightly organized arrangement of lipids, proteins, and protein-sugar compounds forms the plasma membrane, or outer boundary, of certain cells. The organelles, membrane-bound compartments in cells, are built largely from proteins. Biochemical reactions in cells are guided by enzymes, specialized proteins that speed up chemical reactions. The nucleic acid deoxyribonucleic acid (DNA) contains the hereditary information for cells, and another nucleic acid, ribonucleic acid(RNA), works with DNA to build the thousands of proteins the cell needs.

CELL STRUCTURE

Cells fall into one of two categories: prokaryotic or eukaryotic. In a prokaryotic cell, found only in bacteria and archaebacteria, all the components, including the DNA, mingle freely in the cell's interior, a single compartment. Eukaryotic cells, which make up plants, animals, fungi, and all other life forms, contain numerous compartments, or organelles, within each cell. The DNA in eukaryotic cells is enclosed in a special organelle called the nucleus, which serves as the cell's command centre and information library. The term *prokaryote* comes from Greek words that mean "before nucleus" or "prenucleus," while *eukaryote* means "true nucleus."

Prokaryotic Cells

Prokaryotic cells are among the tiniest of all cells, ranging in size from 0.0001 to 0.003 mm (0.000004 to 0.0001 in) in diameter. About a hundred typical prokaryotic cells lined up in a row would match the thickness of a book page. These cells, which can be rodlike, spherical, or spiral in shape, are surrounded by a protective cell wall.

Like most cells, prokaryotic cells live in a watery environment, whether it is soil moisture, a pond, or the fluid surrounding cells in the human body. Tiny pores in the cell wall enable water and the substances dissolved in it, such as oxygen, to flow into the cell; these pores also allow wastes to flow out.

Pushed up against the inner surface of the prokaryotic cell wall is a thin membrane called the plasma membrane. The plasma membrane, composed of two layers of flexible lipid molecules and interspersed with durable proteins, is both supple and strong. Unlike the cell wall, whose open pores allow the unregulated traffic of materials in and out of the cell, the plasma membrane is selectively permeable, meaning it allows only certain substances to pass through. Thus, the plasma membrane actively separates the cell's contents from its surrounding fluids.

While small molecules such as water, oxygen, and carbon dioxide diffuse freely across the plasma membrane, the passage of many larger molecules, including amino acids (the building blocks of proteins) and sugars, is carefully regulated. Specialized transport proteins accomplish this task. The transport proteins span the plasma membrane, forming an intricate system of pumps and channels through which traffic is conducted. Some substances swirling in the fluid around the cell can enter it only if they bind to and are escorted in by specific transport proteins. In this way, the cell fine-tunes its internal environment.

The plasma membrane encloses the cytoplasm, the semifluid that fills the cell. Composed of about 65 percent water, the cytoplasm is packed with up to a billion molecules per cell, a rich storehouse that includes enzymes and dissolved nutrients, such as sugars and amino acids. The water provides a favorable environment for the thousands of biochemical reactions that take place in the cell.

Within the cytoplasm of all prokaryotes is deoxyribonucleic acid (DNA), a complex molecule in the form of a double helix, a shape similar to a spiral staircase. The DNA is about 1,000 times the length of the cell, and to fit inside, it repeatedly twists and folds to form a compact structure called a chromosome. The chromosome in prokaryotes is circular, and is located in a region of the cell called the nucleoid.

Often, smaller chromosomes called plasmids are located in the cytoplasm. The DNA is divided into units called genes, just like a long train is divided into separate cars. Depending on the species, the DNA contains several hundred or even thousands of genes. Typically, one gene contains coded instructions for building all or part of a single protein. Enzymes, which are specialized proteins, determine virtually all the biochemical reactions that support and sustain the cell.

Also immersed in the cytoplasm are the only organelles in prokaryotic cells—tiny bead-like structures called ribosomes. These are the cell's protein factories. Following the instructions encoded in the DNA, ribosomes churn out proteins by the hundreds every minute, providing needed enzymes, the replacements for worn-out transport proteins, or other proteins required by the cell.

While relatively simple in construction, prokaryotic cells display extremely complex activity. They have a greater range of biochemical reactions than those found in their larger relatives, the eukaryotic cells. The extraordinary biochemical diversity of prokaryotic cells is manifested in the wide-ranging lifestyles of the archaebacteria and the bacteria, whose habitats include polar ice, deserts, and hydrothermal vents—deep regions of the ocean under great pressure where hot water geysers erupt from cracks in the ocean floor.

Eukaryotic Animal Cells

Eukaryotic cells are typically about ten times larger than prokaryotic cells. In animal cells, the plasma membrane, rather than a cell wall, forms the cell's outer boundary. With a design similar to the plasma membrane of prokaryotic cells, it separates the cell from its surroundings and regulates the traffic across the membrane.

The eukaryotic cell cytoplasm is similar to that of the prokaryote cell except for one major difference: Eukaryotic cells house a nucleus and numerous other membrane-enclosed organelles. Like separate rooms of a house, these organelles enable specialized functions to be carried out efficiently. The building of proteins and lipids, takes place in separate organelles where specialized enzymes geared for each job are located.

The nucleus is the largest organelle in an animal cell. It contains numerous strands of DNA, the length of each strand being many times the diameter of the cell. Unlike the circular prokaryotic DNA, long sections of eukaryotic DNA pack into the nucleus by wrapping around proteins. As a cell begins to divide, each DNA strand folds over onto itself several times, forming a rod-shaped chromosome.

The nucleus is surrounded by a double-layered membrane that protects the DNA from potentially damaging chemical reactions that occur in the cytoplasm. Messages pass between the cytoplasm and the nucleus through nuclear pores, which are holes in the membrane of the nucleus. In each nuclear pore, molecular signals flash back and forth as often as ten times per second. A signal to activate a specific gene comes in to the nucleus and instructions for production of the necessary protein go out to the cytoplasm.

Attached to the nuclear membrane is an elongated membranous sac called the endoplasmic reticulum. This organelle tunnels through the cytoplasm, folding back and forth on itself to form a series of membranous stacks. Endoplasmic reticulum takes two forms: rough and smooth. Rough endoplasmic reticulum (RER) is so called because it appears bumpy under a microscope. The bumps are actually thousands of ribosomes attached to the membrane's surface. The ribosomes in eukaryotic cells have the same function as those in prokaryotic cells—protein synthesis—but they differ slightly in structure.

Eukaryote ribosomes bound to the endoplasmic reticulum help assemble proteins that typically are exported from the cell. The ribosomes work with other molecules to link amino acids to partially completed proteins. These incomplete proteins then travel to the inner chamber of the endoplasmic reticulum, where chemical modifications, such as the addition of a sugar, are carried out. Chemical modifications of lipids are also carried out in the endoplasmic reticulum.

The endoplasmic reticulum and its bound ribosomes are particularly dense in cells that produce many proteins for export, such as the white blood cells of the immune system, which produce and secrete antibodies. Some ribosomes that manufacture proteins are not attached to the endoplasmic reticulum. These so-called free ribosomes are dispersed in the cytoplasm and typically make proteins—many of them enzymes—that remain in the cell.

The second form of endoplasmic reticulum, the smooth endoplasmic reticulum (SER), lacks ribosomes and has an even surface. Within the winding channels of the smooth endoplasmic reticulum are the enzymes needed for the construction of molecules such as carbohydrates and lipids. The smooth endoplasmic reticulum is prominent in liver cells, where it also serves to detoxify substances such as alcohol, drugs, and other poisons.

Proteins are transported from free and bound ribosomes to the Golgi apparatus, an organelle that resembles a stack of deflated balloons. It is packed with enzymes that complete the processing of proteins.

These enzymes add sulfur or phosphorus atoms to certain regions of the protein, or chop off tiny pieces from the ends of the proteins. The completed protein then leaves the Golgi apparatus for its final destination inside or outside the cell. During its assembly on the ribosome, each protein has acquired a group of from 4 to 100 amino acids called a signal. The signal works as a molecular shipping label to direct the protein to its proper location.

Lysosomes are small, often spherical organelles that function as the cell's recycling centre and garbage disposal. Powerful digestive enzymes concentrated in the lysosome break down worn-out organelles and ship their building blocks to the cytoplasm where they are used to construct new organelles. Lysosomes also dismantle and recycle proteins, lipids, and other molecules.

The mitochondria are the powerhouses of the cell. Within these long, slender organelles, which can appear oval or bean shaped under the electron microscope, enzymes convert the sugar glucose and other nutrients into adenosine triphosphate (ATP). This molecule, in turn, serves as an energy battery for countless cellular processes, including the shuttling of substances across the plasma membrane, the building and transport of proteins and lipids, the recycling of molecules and organelles, and the dividing of cells.

Muscle and liver cells are particularly active and require dozens and sometimes up to a hundred mitochondria per cell to meet their energy needs. Mitochondria are unusual in that they contain their own DNA in the form of a prokaryote-like circular chromosome; have their own ribosomes, which resemble prokaryotic ribosomes; and divide independently of the cell.

Unlike the tiny prokaryotic cell, the relatively large eukaryotic cell requires structural support.

The cytoskeleton, a dynamic network of protein tubes, filaments, and fibres, crisscrosses the cytoplasm, anchoring the organelles in place and providing shape and structure to the cell. Many components of the cytoskeleton are assembled and disassembled by the cell as needed. During cell division, a special structure called a spindle is built to move chromosomes around. After cell division, the spindle, no longer needed, is dismantled. Some components of the cytoskeleton serve as microscopic tracks along which proteins and other molecules travel like miniature trains. Recent research suggests that the cytoskeleton also may be a mechanical communication structure that converses with the nucleus to help organize events in the cell.

Eukaryotic Plant Cells

Plant cells have all the components of animal cells and boast several added features, including chloroplasts, a central vacuole, and a cell wall. Chloroplasts convert light energy—typically from the Sun—into the sugar glucose, a form of chemical energy, in a process known as photosynthesis. Chloroplasts, like mitochondria, possess a circular chromosome and prokaryote-like ribosomes, which manufacture the proteins that the chloroplasts typically need.

The central vacuole of a mature plant cell typically takes up most of the room in the cell. The vacuole, a membranous bag, crowds the cytoplasm and organelles to the edges of the cell. The central vacuole stores water, salts, sugars, proteins, and other nutrients. In addition, it stores the blue, red, and purple pigments that give certain flowers their colours. The central vacuole also contains plant wastes that taste bitter to certain insects, thus discouraging the insects from feasting on the plant.

In plant cells, a sturdy cell wall surrounds and protects the plasma membrane. Its pores enable materials to pass freely into and out of the cell. The strength of the wall also enables a cell to absorb water into the central vacuole and swell without bursting. The resulting pressure in the cells provides plants with rigidity and support for stems, leaves, and flowers. Without sufficient water pressure, the cells collapse and the plant wilts.

CELL FUNCTIONS

To stay alive, cells must be able to carry out a variety of functions. Some cells must be able to move, and most cells must be able to divide. All cells must maintain the right concentration of chemicals in their cytoplasm, ingest food and use it for energy, recycle molecules, expel wastes, and construct proteins. Cells must also be able to respond to changes in their environment.

Movement

Many unicellular organisms swim, glide, thrash, or crawl to search for food and escape enemies. Swimming organisms often move by means of a flagellum, a long tail-like structure made of protein. Many bacteria, have one, two, or many flagella that rotate like propellers to drive the organism along. Some single-celled eukaryotic organisms, such as euglena, also have a flagellum, but it is longer and thicker than the prokaryotic flagellum.

The eukaryotic flagellum works by waving up and down like a whip. In higher animals, the sperm cell uses a flagellum to swim toward the female egg for fertilization.

Movement in eukaryotes is also accomplished with cilia, short, hairlike proteins built by centrioles, which are barrel-shaped structures located in the cytoplasm that assemble and break down protein filaments. Typically, thousands of cilia extend through the plasma membrane and cover the surface of the cell, giving it a dense, hairy appearance. By beating its cilia as if they were oars, an organism such as the paramecium propels itself through its watery environment. In cells that do not move, cilia are used for other purposes. In the respiratory tract of humans, millions of ciliated cells prevent inhaled dust, smog, and microorganisms from entering the lungs by sweeping them up on a current of mucus into the throat, where they are swallowed. Eukaryotic flagella and cilia are formed from basal bodies, small protein structures located just inside the plasma membrane. Basal bodies also help to anchor flagella and cilia.

Still other eukaryotic cells, such as amoebas and white blood cells, move by amoeboid motion, or crawling. They extrude their cytoplasm to form temporary pseudopodia, or false feet, which actually are placed in front of the cell, rather like extended arms. They then drag the trailing end of their cytoplasm up to the pseudopodia. A cell using amoeboid motion would lose a race to a euglena or paramecium. But while it is slow, amoeboid motion is strong enough to move cells against a current, enabling water-dwelling organisms to pursue and devour prey, or white blood cells roaming the blood stream to stalk and engulf a bacterium or virus.

Nutrition

All cells require nutrients for energy, and they display a variety of methods for ingesting them. Simple nutrients dissolved in pond water, can be carried through the plasma membrane of pond-dwelling organisms via a series of molecular pumps. In humans, the cavity of the small intestine contains the nutrients from digested food, and cells that form the walls of the intestine use similar pumps to pull amino acids and other nutrients from the cavity into the bloodstream. Certain unicellular organisms, such as amoebas, are also capable of reaching out and grabbing food. They use a process known as endocytosis, in which the plasma membrane surrounds and engulfs the food particle, enclosing it in a sac, called a vesicle, that is within the amoeba's interior.

Energy

Cells require energy for a variety of functions, including moving, building up and breaking down molecules, and transporting substances across the plasma membrane. Nutrients contains energy, but cells must convert the energy locked in nutrients to another form—specifically, the ATP molecule, the cell's energy battery—before it is useful. In single-celled eukaryotic organisms, such as the paramecium, and in multicellular eukaryotic organisms, such as plants, animals, and fungi, mitochondria are responsible for this task.

The interior of each mitochondrion consists of an inner membrane that is folded into a mazelike arrangement of separate compartments called cristae. Within the cristae, enzymes form an assembly line where the energy in glucose and other energy-rich nutrients is harnessed to build ATP; thousands of ATP molecules are constructed each second in a typical cell. In most eukaryotic cells, this process requires oxygen and is known as aerobic respiration.

Some prokaryotic organisms also carry out aerobic respiration. They lack mitochondria, however, and carry out aerobic respiration in the cytoplasm with the help of enzymes sequestered there. Many prokaryote species live in environments where there is little or no oxygen, environments such as mud, stagnant ponds, or within the intestines of animals. Some of these organisms produce ATP without oxygen in a process known as anaerobic respiration, where sulfur or other substances take the place of oxygen. Still other prokaryotes, and yeast, a single-celled eukaryote, build ATP without oxygen in a process known as fermentation.

Almost all organisms rely on the sugar glucose to produce ATP. Glucose is made by the process of photosynthesis, in which light energy is transformed to the chemical energy of glucose. Animals and fungi cannot carry out photosynthesis and depend on plants and other photosynthetic organisms for this task. In plants, as we have seen, photosynthesis takes place in organelles called chloroplasts.

Chloroplasts contain numerous internal compartments called thylakoids where enzymes aid in the energy conversion process. A single leaf cell contains 40 to 50 chloroplasts. With sufficient sunlight, one large tree is capable of producing upwards of two tons of sugar in a single day. Photosynthesis in prokaryotic organisms—typically aquatic bacteria—is carried out with enzymes clustered in plasma membrane folds called chromatophores. Aquatic bacteria produce the food consumed by tiny organisms living in ponds, rivers, lakes, and seas.

Protein Synthesis

A typical cell must have on hand about 30,000 proteins at any one time. Many of these proteins are enzymes needed to construct the major molecules used by cells—carbohydrates, lipids, proteins, and nucleic acids—or to aid in the breakdown of such molecules after they have worn out. Other proteins are part of the cell's structure—the plasma membrane and ribosomes. In animals, proteins also function as hormones and antibodies, and they function like delivery trucks to transport other molecules around the body. Hemoglobin, is a protein that transports oxygen in red blood cells. The cell's demand for proteins never ceases.

Before a protein can be made, however, the molecular directions to build it must be extracted from one or more genes. In humans, one gene holds the information for the protein insulin, the hormone that cells need to import glucose from the bloodstream, while at least two genes hold the information for collagen, the protein that imparts strength to skin, tendons, and ligaments.

The process of building proteins begins when enzymes, in response to a signal from the cell, bind to the gene that carries the code for the required protein, or part of the protein. The enzymes transfer the code to a new molecule called messenger RNA, which carries the code from the nucleus to the cytoplasm. This enables the original genetic code to remain safe in the nucleus, with messenger RNA delivering small bits and pieces of information from the DNA to the cytoplasm as needed. Depending on the cell type, hundreds or even thousands of molecules of messenger RNA are produced each minute.

Once in the cytoplasm, the messenger RNA molecule links up with a ribosome. The ribosome moves along the messenger RNA like a monorail car along a track, stimulating another form of RNA—transfer RNA—to gather and link the necessary amino acids, pooled in the cytoplasm, to form the specific protein, or section of protein. The protein is modified as necessary by the endoplasmic reticulum and Golgi apparatus before embarking on its mission. Cells teem with activity as they forge the numerous, diverse proteins that are indispensable for life.

Cell Division

Most cells divide at some time during their life cycle, and some divide dozens of times before they die. Organisms rely on cell division for reproduction, growth, and repair and replacement of damaged or worn out cells. Three types of cell division occur: binary fission, mitosis, and meiosis. Binary fission, the method used by prokaryotes, produces two identical cells from one cell. The more complex process of mitosis, which also produces two genetically identical cells from a single cell, is used by many unicellular eukaryotic organisms for reproduction.

Multicellular organisms use mitosis for growth, cell repair, and cell replacement. In the human body, an estimated 25 million mitotic cell divisions occur every second in order to replace cells that have completed their normal life cycles. Cells of the liver, intestine, and skin may be replaced every few days. Recent research indicates that even brain cells, once thought to be incapable of mitosis, undergo cell division in the part of the brain associated with memory.

The type of cell division required for sexual reproduction is meiosis. Sexually reproducing organisms include seaweeds, fungi, plants, and animals—including, of course, human beings. Meiosis differs from mitosis in that cell division begins with a cell that has a full complement of chromosomes and ends with gamete cells, such as sperm and eggs, that have only half the complement of chromosomes. When a sperm and egg unite during fertilization, the cell resulting from the union, called a zygote, contains the full number of chromosomes.

ORIGIN OF CELLS

The story of how cells evolved remains an open and actively investigated question in science The combined expertise of physicists, geologists, chemists, and evolutionary biologists has been required to shed light on the evolution of cells from the nonliving matter of early Earth.

The planet formed about 4.5 billion years ago, and for millions of years, violent volcanic eruptions blasted substances such as carbon dioxide, nitrogen, water, and other small molecules into the air.

These small molecules, bombarded by ultraviolet radiation and lightning from intense storms, collided to form the stable chemical bonds of larger molecules, such as amino acids and nucleotides—the building blocks of proteins and nucleic acids. Experiments indicate that these larger molecules form spontaneously under laboratory conditions that simulate the probable early environment of Earth.

Scientists speculate that rain may have carried these molecules into lakes to create a primordial soup—a breeding ground for the assembly of proteins, the nucleic acid RNA, and lipids. Some scientists postulate that these more complex molecules formed in hydrothermal vents rather than in lakes. Other scientists propose that these key substances may have reached Earth on meteorites from outer space. Regardless of the origin or environment, however, scientists do agree that proteins, nucleic acids, and lipids provided the raw materials for the first cells.

In the laboratory, scientists have observed lipid molecules joining to form spheres that resemble a cell's plasma membrane. As a result of these observations, scientists postulate that millions of years of molecular collisions resulted in lipid spheres enclosing RNA, the simplest molecule capable of self-replication. These primitive aggregations would have been the ancestors of the first prokaryotic cells.

Fossil studies indicate that cyanobacteria, bacteria capable of photosynthesis, were among the earliest bacteria to evolve, an estimated 3.4 billion to 3.5 billion years ago. In the environment of the early Earth, there was no oxygen, and cyanobacteria probably used fermentation to produce ATP. Over the eons, cyanobacteria performed photosynthesis, which produces oxygen as a byproduct; the result was the gradual accumulation of oxygen in the atmosphere. The presence of oxygen set the stage for the evolution of bacteria that used oxygen in aerobic respiration, a more efficient ATP-producing process than fermentation.

Some molecular studies of the evolution of genes in archaebacteria suggest that these organisms may have evolved in the hot waters of hydrothermal vents or hot springs slightly earlier than cyanobacteria, around 3.5 billion years ago. Like cyanobacteria, archaebacteria probably relied on fermentation to synthesize ATP.

Eukaryotic cells may have evolved from primitive prokaryotes about 2 billion years ago. One hypothesis suggests that some prokaryotic cells lost their cell walls, permitting the cell's plasma membrane to expand and fold. These folds, ultimately, may have given rise to separate compartments within the cell—the forerunners of the nucleus and other organelles now found in eukaryotic cells. Another key hypothesis is known as endosymbiosis.

Molecular studies of the bacteria-like DNA and ribosomes in mitochondria and chloroplasts indicate that mitochondrion and chloroplast ancestors were once free-living bacteria. Scientists propose that these free-living bacteria were engulfed and maintained by other prokaryotic cells for their ability to produce ATP efficiently and to provide a steady supply of glucose. Over generations, eukaryotic cells complete with mitochondria—the ancestors of animals—or with both mitochondria and chloroplasts—the ancestors of plants—evolved.

THE DISCOVERY AND STUDY OF CELLS

The first observations of cells were made in 1665 by English scientist Robert Hooke, who used a crude microscope of his own invention to examine a variety of objects, including a thin piece of cork. Noting the rows of tiny boxes that made up the dead wood's tissue, Hooke coined the term *cell* because the boxes reminded him of the small cells occupied by monks in a monastery. While Hooke was the first to observe and describe cells, he did not comprehend their significance. At about the same time, the Dutch maker of microscopes Antoni van Leeuwenhoek pioneered the invention of one of the best microscopes of the time. Using his invention, Leeuwenhoek was the first to observe, draw, and describe a variety of living organisms, including bacteria gliding in saliva, one-celled organisms cavorting in pond water, and sperm swimming in semen. Two centuries passed, however, before scientists grasped the true importance of cells.

Modern ideas about cells appeared in the 1800s, when improved light microscopes enabled scientists to observe more details of cells. Working together, German botanist Matthias Jakob Schleiden and German zoologist Theodor Schwann recognized the fundamental similarities between plant and animal cells. In 1839 they proposed the revolutionary idea that all living things are made up of cells. Their theory gave rise to modern biology: a whole new way of seeing and investigating the natural world.

By the late 1800s, as light microscopes improved still further, scientists were able to observe chromosomes within the cell. Their research was aided by new techniques for staining parts of the cell, which made possible the first detailed observations of cell division, including observations of the differences between mitosis and meiosis in the 1880s. In the first few decades of the 20th century, many scientists focused on the behaviour of chromosomes during cell division. At that time, it was generally held that mitochondria transmitted the hereditary information. By 1920, however, scientists determined that chromosomes carry genes and that genes transmit hereditary information from generation to generation.

During the same period, scientists began to understand some of the chemical processes in cells. In the 1920s, the ultracentrifuge was developed. The ultracentrifuge is an instrument that spins celis or other substances in test tubes at high speeds, which causes the heavier parts of the substance to fall to the bottom of the test tube. This instrument enabled scientists to separate the relatively abundant and heavy mitochondria from the rest of the cell and study their chemical reactions. By the late 1940s, scientists were able to explain the role of mitochondria in the cell. Using refined techniques with the ultracentrifuge, scientists subsequently isolated the smaller organelles and gained an understanding of their functions.

While some scientists were studying the functions of cells, others were examining details of their structure. They were aided by a crucial technological development in the 1940s: the invention of the electron microscope, which uses high-energy electrons instead of light waves to view specimens. New generations of electron microscopes have provided resolution, or the differentiation of separate objects, thousands of times more powerful than that available in light microscopes. This powerful resolution revealed organelles such as the endoplasmic reticulum, lysosomes, the Golgi apparatus, and the cytoskeleton. The scientific fields of cell structure and function continue to complement each other as scientists explore the enormous complexity of cells.

The discovery of the structure of DNA in 1953 by American biochemist James D. Watson and British biophysicist Francis Crick ushered in the era of molecular biology. Today, investigation inside the world of cells—of genes and proteins at the molecular level—constitutes one of the largest and fastest moving areas in all of science. One particularly active field in recent years has been the investigation of cell signaling, the process by which molecular messages find their way into the cell via a series of complex protein pathways in the cell.

Another busy area in cell biology concerns programmed cell death, or apoptosis. Millions of times per second in the human body, cells commit suicide as an essential part of the normal cycle of cellular replacement. This also seems to be a check against disease: When mutations build up within a cell, the cell will usually self-destruct. If this fails to occur, the cell may divide and give rise to mutated daughter cells, which continue to divide and spread, gradually forming a growth called a tumor. This unregulated growth by rogue cells can be benign, or harmless, or cancerous, which may threaten healthy tissue. The study of apoptosis is one avenue that scientists explore in an effort to understand how cells become cancerous.

Scientists are also discovering exciting aspects of the physical forces within cells. Cells employ a form of architecture called tensegrity, which enables them to withstand battering by a variety of mechanical stresses, such as the pressure of blood flowing around cells or the movement of organelles within the cell. Tensegrity stabilizes cells by evenly distributing mechanical stresses to the cytoskeleton and other cell components.

Tensegrity also may explain how a change in the cytoskeleton, where certain enzymes are anchored, initiates biochemical reactions within the cell, and can even influence the action of genes. The mechanical rules of tensegrity may also account for the assembly of molecules into the first cells. Such new insights—made some 300 years after the tiny universe of cells was first glimpsed—show that cells continue to yield fascinating new worlds of discovery.

RIBOSOME

Ribosome, cell structure that uses genetic instructions transported in ribonucleic acid (RNA) to link a specific sequence of amino acids into chains to form proteins. Ribosomes, which measure about 0.00025 mm (0.00001 in), are dispersed in the cytoplasm (the cell contents outside the nucleus) of all prokaryotic cells— archaebacteria and bacteria. They are also found in the cytoplasm of all eukaryotic cells—cells of protists, fungi, plants, and animals—where they either float free in the cytoplasm or are bound to networks of membrane-enclosed tubules in the cytoplasm, called the endoplasmic reticulum. In eukaryotic cells, two types of cell structures called mitochondria and chloroplasts also contain ribosomes.

Structure

Development of electron microscope techniques in the 1950s enabled scientists to detect tiny granules in the cell's cytoplasm. By separating the granules from the rest of the cell, scientists were able to study their composition. Because the granules are rich in ribonucleic acid (RNA), scientists named them ribosomes.

Ribosomes in the cytoplasm of eukaryotic cells differ slightly from ribosomes in prokaryotic cells. Eukaryotic ribosomes consist of four strands of RNA associated with 70 to 80 proteins. In contrast, prokaryotic ribosomes each contain only three strands of RNA and about 50 proteins. Prokaryotic ribosomes are smaller and less dense than eukaryotic ribosomes. The ribosomes located in the mitochondria and chloroplasts of eukaryotic cells more closely resemble prokaryotic ribosomes, a finding that suggests mitochondria and chloroplasts evolved from once free-living prokaryotes.

Synthesis

Ribosomes are required for protein synthesis; when cells need large numbers of proteins, they must first build numerous—sometimes hundreds of thousands—of ribosomes. To ensure rapid production of the RNA portion of ribosomes, certain eukaryotic cells contain thousands of copies of genes that code for ribosomal RNA. The locations of all the genes that code for the protein component of ribosomes are yet to be determined.

In eukaryotic cells, three of the four ribosomal RNA strands are synthesized in the nucleolus, a dense, granular structure in the nucleus. The fourth ribosomal RNA strand is synthesized outside of the nucleolus and then transported into the nucleolus for ribosome assembly.

Ribosomal proteins enter the nucleolus and combine with the four ribosomal RNA strands to create two ribosomal structures: the large and small subunits. These two beadlike subunits leave the nucleus separately through special openings called nuclear pores. The two subunits unite outside of the nucleus just before the ribosome begins to manufacture proteins. Prokaryotic cells lack a nucleus or nucleolus, and ribosomal synthesis takes place in the cytoplasm. In these cells, only a few copies of genes coding for ribosomal RNA are present.

Protein Synthesis

Protein synthesis in a cell begins with initiation, when a chain of messenger RNA (mRNA) carrying genetic instructions from deoxyribonucleic acid (DNA) attaches to a ribosome. The mRNA instructs the ribosome how to assemble amino acids to form a protein. Two molecules of transfer RNA (tRNA), each carrying an amino acid, join the ribosome-messenger RNA complex at two positions called P-site and A-site. A chemical bond known as a peptide bond is formed between the first two amino acids.

During elongation the tRNA in the P-site detaches from its amino acid and floats away from the complex while the tRNA carrying the two bonded amino acids moves from the A-site over to the P-site. This leaves the A-site open for a new tRNA molecule carrying a third amino acid to attach to the ribosome. The new amino acid bonds to the second amino acid with another peptide bond. Again, one tRNA is released and the remaining tRNA molecule, now carrying a chain of three amino acids, moves over to the P-site. The ribosome coordinates this cycle repeatedly until termination occurs, when the ribosome encounters a stop signal on the mRNA. The completed protein, which may be a chain of hundreds of amino acids, is released from the ribosome.

In general, DNA carries the genetic instructions for making all cellular structures. Since all cells contain ribosomes, scientists are comparing DNA instructions for making ribosomes in different species to learn how closely the species are related.

MITOCHONDRIA

Mitochondria, small cellular structures, or organelles, found in the cytoplasm of eukaryotic cells (cells with a nucleus). Mitochondria are responsible for converting nutrients into the energy-yielding molecule adenosine triphosphate (ATP) to fuel the cell's activities. This function, known as aerobic respiration, is the reason mitochondria are frequently referred to as the powerhouse of the cell.

Mitochondria are unusual organelles in that they contain deoxyribonucleic acid (DNA), typically found in the cell's nucleus, and ribosomes, protein-producing organelles abundant in the cytoplasm. Within the mitochondria, the DNA directs the ribosomes to produce proteins, many of which function as enzymes, or biological catalysts, in ATP production. The number of mitochondria in a cell depends on the cell's function. Cells with particularly heavy energy demands, such as muscle cells, have more mitochondria than other cells.

Mitochondrial Structure

A mitochondrion is typically long and slender, but it can appear bean-shaped or oval-shaped under the electron microscope. Ranging in size from 0.5 micrometer (0.00005 in) to 1 micrometer (0.0001 in) in length, a mitochondrion has a double membrane that forms a sac within a sac. The smooth outer membrane holds numerous transport proteins, which shuttle materials in and out of the mitochondrion. The region between the outer and inner membranes, which is filled with liquid, is known as the outer compartment.

The inner membrane has numerous folds called *cristae*. Cristae are the sites of ATP synthesis, and their folded structure greatly increases the surface area where ATP synthesis occurs. Transport proteins, molecules called electron transport chains, and enzymes that synthesize ATP are among the molecules embedded in the cristae. The cristae enclose a liquid-filled region known as the inner compartment, or matrix, which contains a large number of enzymes that are used in the process of aerobic respiration.

Mitochondrial Function

The chief function of the mitochondria is to create energy for cellular activity by the process of aerobic respiration. In this process, glucose is broken down in the cell's cytoplasm to form pyruvic acid, which is transported into the mitochondrion. In a series of reactions, part of which is called the citric acid cycle or Krebs cycle, the pyruvic acid reacts with water to produce carbon dioxide and ten hydrogen atoms. These hydrogen atoms are transported on special carrier molecules called coenzymes to the cristae, where they are donated to the electron transport chain.

The electron transport chain separates the electron and proton in each of the ten hydrogen atoms. The ten electrons are sent through the electron transport chain and some eventually combine with oxygen and the protons to form water.

Energy is released as the electrons flow from the coenzymes down the electron transport chain to the oxygen atoms, and this energy is trapped by the components of the electron transport chain. As the electrons flow from one component to another, the components pump random protons from the matrix to the outer compartment. The protons cannot return to the matrix except by one pathway—through the enzyme ATPase, which is embedded in the inner membrane. As the protons flow back into the matrix, ATPase adds a phosphate group to a molecule in the matrix, adenosine diphosphate (ADP). The addition of a phosphate group to ADP forms ATP.

Aerobic respiration is an ongoing process, and mitochondria can produce hundreds of thousands of ATP molecules each minute in a typical cell. The ATP is transported to the cytoplasm of the cell, where it is used for virtually every energy-requiring reaction it performs. As ATP is used, it is converted into ADP, which is returned by the cell to the mitochondrion and is used to build more ATP.

Origin of Mitochondria

Mitochondria have significant features that resemble those of prokaryotes, primitive cells that lack a nucleus. Mitochondrial DNA is circular, like the DNA of prokaryotes, and mitochondrial ribosomes are similar to prokaryotic ribosomes. Mitochondria divide independently of the cell through binary fission, the method of cell division typical of prokaryotes.

The prokaryote-like features of mitochondria lead many scientists to support the endosymbiosis hypothesis. This hypothesis states that millions of years ago, free-living prokaryotes capable of aerobic respiration were engulfed by other, larger prokaryotes but not digested, possibly because they were able to resist digestive enzymes. The two cells developed a symbiotic, or cooperative, relationship in which the host cell provided nutrients and the engulfed cell used these nutrients to carry out aerobic respiration, which provided the host cell with an abundant supply of ATP. The engulfed cells evolved into mitochondria, which retain the DNA and ribosomes characteristic of their prokaryotic ancestors.

CHLOROPLAST

Chloroplast, structure in the cells of plants and algae where photosynthesis takes place. Chloroplasts are mostly disk-shaped organelles, 4 to 6 micrometers in diameter. They occur most abundantly in leaf cells, where they can apparently orient themselves to light. Perhaps 40 to 50 chloroplasts exist in one cell and 500,000 in each sq mm (0.06 sq in) of leaf surface. Each chloroplast is enclosed in a double membrane. Internally, it consists of a ground substance called the stroma, which is traversed by a complex network of interconnected disks called thylakoids. Many of the thylakoids are stacked like saucers; the stacks are called grana.

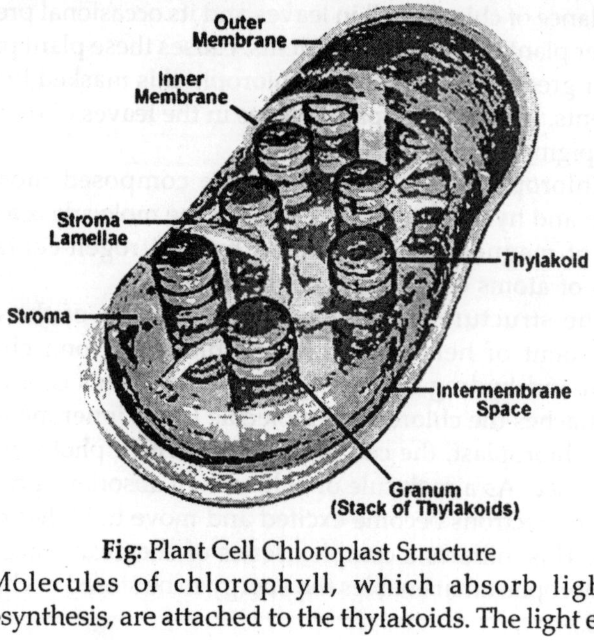

Fig: Plant Cell Chloroplast Structure

Molecules of chlorophyll, which absorb light for photosynthesis, are attached to the thylakoids. The light energy captured by the chlorophyll is converted to adenosine triphosphate, or ATP, in a series of chemical reactions that take place in the grana. Chloroplasts also contain small starch grains that temporarily store the products of photosynthesis.

In plants, chloroplasts develop in the presence of light from small, colourless organelles called proplastids. As cells divide in the growing parts of a plant, the proplastids inside them divide by fission. Thus, the daughter cells have the ability to produce chloroplasts. In algae, chloroplasts divide directly, without developing from proplastids.

The self-reproducing ability of chloroplasts, their bacteria-like DNA and ribosomes, and their close similarity regardless of the type of cell they inhabit, suggest that they were once independent organisms that come to exist in symbiosis with the plant cell as host.

Chlorophyll, the pigment found in plants, some algae, and some bacteria that gives them their green colour and that absorbs the light necessary for photosynthesis. Chlorophyll absorbs mainly violet-blue and orange-red light. The great abundance of chlorophyll in leaves and its occasional presence in other plant tissues, such as stems, causes these plant parts to appear green. In some leaves, chlorophyll is masked by other pigments. In fall, chlorophyll wanes in the leaves of trees, and other pigments predominate.

Chlorophyll is a large molecule composed mostly of carbon and hydrogen. At the centre of the molecule is a single atom of magnesium surrounded by a nitrogen-containing group of atoms called a porphyrin ring.

The structure somewhat resembles that of the active constituent of hemoglobin in the blood. A long chain of carbon and hydrogen atoms proceeds from this central core and attaches the chlorophyll molecule to the inner membrane of the chloroplast, the cell organelle in which photosynthesis takes place. As a molecule of chlorophyll absorbs a photon of light, its electrons become excited and move to higher energy levels. This initiates a complex series of chemical reactions in the chloroplast that enables the energy to be stored in chemical bonds.

Several kinds of chlorophyll exist. They differ from each other in details of their molecular structure and absorb slightly different wavelengths of light. The most common type is chlorophyll a, making up about 75 percent of the chlorophyll in green plants. It is also found in cyanobacteria (formerly known as blue-green algae) and in more complex photosynthetic cells.

Chlorophyll b is an accessory pigment present in plants and other complex photosynthetic cells; it absorbs light energy of a different wavelength and transfers it to chlorophyll a for ultimate conversion to chemical energy. Other chlorophylls, of minor importance, are found in some bacteria.

PARAMECIUM

Paramecium, genus of protozoa of the phylum Ciliophora, often called slipper animalcules because of their slipperlike shape. Paramecia are unicellular organisms usually less than 0.25 mm (0.01 in) in length and covered with minute hairlike projections called cilia. Cilia are used in locomotion and during feeding. When moving through the water, paramecia follow a spiral path while rotating on the long axis.

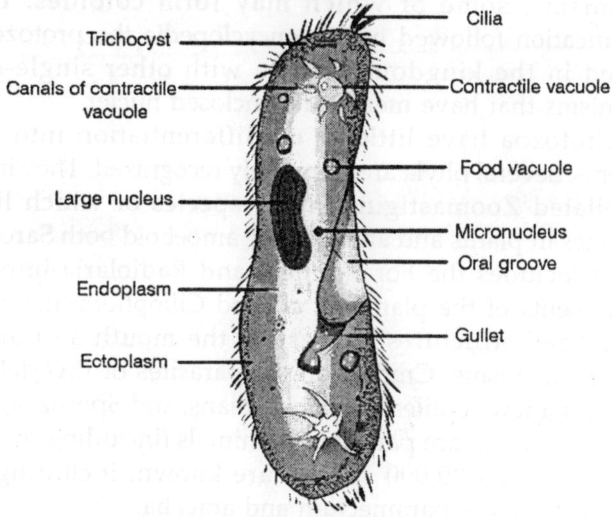

Fig: Structure of a Paramecius

When a paramecium encounters an obstacle, it exhibits the so-called avoidance reaction: It backs away at an angle and starts off in a new direction. Paramecia feed mostly on bacteria, which are driven into the gullet by the cilia. Two contractile vacuoles regulate osmotic pressure and also serve as excretory structures.

A paramecium has a large nucleus called a macronucleus, without which it cannot survive, and one or two small nuclei called micronuclei, without which it cannot reproduce sexually.

Reproduction is usually asexual by transverse binary fission, occasionally sexual by conjugation, and rarely by endomixis, a process involving total nuclear reorganization of individual organisms.

Paramecia abound in freshwater ponds throughout the world; one species lives in marine waters. They are easily cultivated in the laboratory by allowing vegetable matter to stand in water for a few days. The common species *Paramecium caudatum* is widely used in research.

PROTOZOA

Protozoa, collective name for animal-like, single-celled organisms, some of which may form colonies. In the classification followed in this encyclopedia the protozoa are placed in the kingdom Protista with other single-celled organisms that have membrane-enclosed nuclei.

Protozoa have little or no differentiation into tissue systems. Several phyla are commonly recognized. They include flagellated Zoomastigina, many species of which live as parasites in plants and animals; the amoeboid both Sarcodina, which includes the Foraminifera and Radiolaria important components of the plankton; ciliated Ciliophora, many with specialized structures suggesting the mouth and anus of higher organisms; Cnidosporidia, parasites of invertebrates, fish, and a few reptiles and amphibians; and Sporozoa, many species of which are parasites of animals (including humans).

More than 20,000 species are known, including such familiar forms as paramecium and amoeba.

Most species are found in such aquatic habitats as oceans,

lakes, rivers, and ponds. They vary in length from 2 to 70 micrometers. Protozoa obtain their food by ingesting bacteria, waste products of other organisms, algae, or other protozoa. Most species are motile, either by whiplike structures called flagella, hairlike structures called cilia, or amoeboid motion, a streaming type of movement involving the formation of pseudopods (footlike extensions).

Chapter 2

Morphology of Flowering Plants and Root

The wide range in the structure of higher plants will never fail to fascinate us. Even though the angiosperms show such a large diversity in external structure or morphology, they are all characterised by presence of roots, stems, leaves, flowers and fruits. Classification of plants based on morphological and other characteristics. For any successful attempt at classification and at understanding any higher plant (or for that matter any living organism) we need to know standard technical terms and standard definitions.

We also need to know about the possible variations in different parts, found as adaptations of the plants to their environment, e.g., adaptions to various habitats, for protection, climbing, storage, etc. If you pull out any weed you will see that all of them have roots, stems and leaves. They may be bearing flowers and fruits. The underground part of the flowering plant is the root system while the portion above the ground forms the shoot system.

THE ROOT

In majority of the dicotyledonous plants, the direct elongation of the radicle leads to the formation of primary root which grows inside the soil. It bears lateral roots of several orders that are referred to as secondary, tertiary, etc. roots.

The primary roots and its branches constitute the tap root system, as seen in the mustard plant. In monocotyledonous plants, the primary root is short lived and is replaced by a large number of roots. These roots originate from the base of the stem and constitute the fibrous root system, as seen in the wheat plant. In some plants, like grass, *Monstera* and the banyan tree, roots arise from parts of the plant other than the radicle and are called adventitious roots. The main functions of the root system are absorption of water and minerals from the soil, providing a proper anchorage to the plant parts, storing reserve food material and synthesis of plant growth regulators.

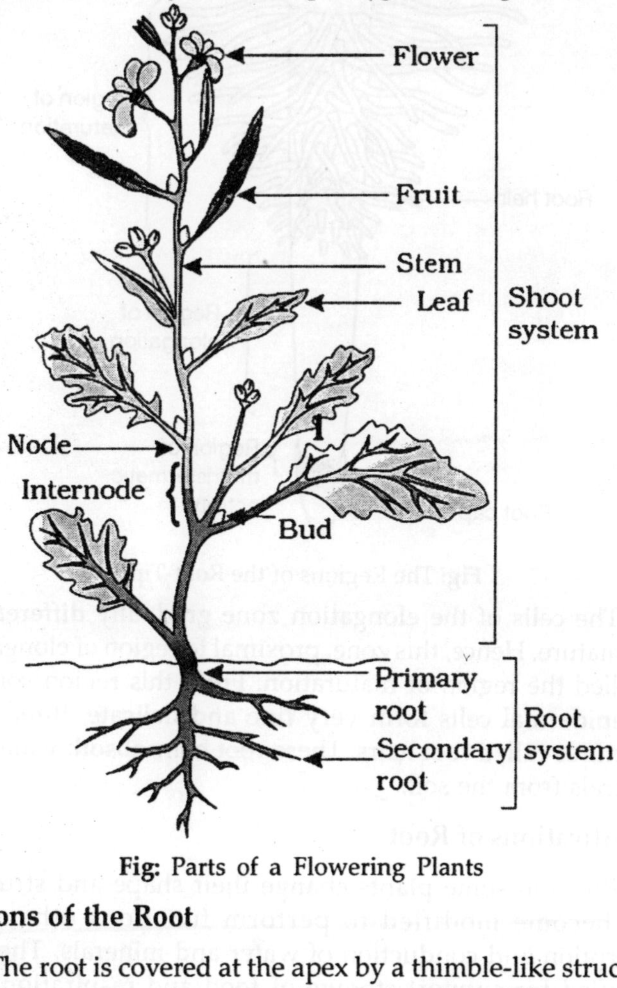

Fig: Parts of a Flowering Plants

Regions of the Root

The root is covered at the apex by a thimble-like structure called the root cap. It protects the tender apex of the root as it makes its way through the soil. A few millimetres above the root cap is the region of meristematic activity. The cells of this region are very small, thin-walled and with dense protoplasm. They divide repeatedly. The cells proximal to this region undergo rapid elongation and enlargement and are responsible for the growth of the root in length. This region is called the region of elongation.

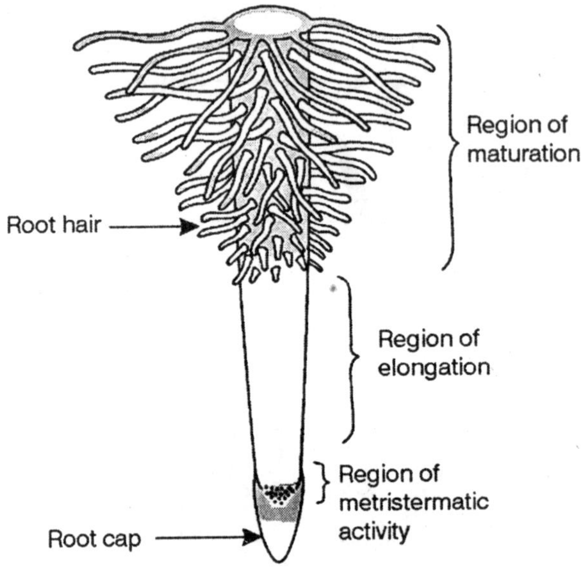

Fig: The Regions of the Root-Tip

The cells of the elongation zone gradually differentiate and mature. Hence, this zone, proximal to region of elongation, is called the region of maturation. From this region some of the epidermal cells form very fine and delicate, thread-like structures called root hairs. These root hairs absorb water and minerals from the soil.

Modifications of Root

Roots in some plants change their shape and structure and become modified to perform functions other than absorption and conduction of water and minerals. They are modified for support storage of food and respiration. Tap roots of carrot, turnips and adventitious roots of sweet potato, get swollen and store food.

Can you give some more such examples? Have you ever wondered what those hanging structures are that support a banyan tree? These are called prop roots. Similarly, the stems of maize and sugarcane have supporting roots coming out of the lower nodes of the stem. These are called stilt roots. In some plants such as *Rhizophora* growing in swampy areas, many roots come out of the ground and grow vertically upwards. Such roots, called pneumatophores, help to get oxygen for respiration.

COMPARATIVE MORPHOLOGY OF ROOTS AND ROOT SYSTEMS

The root system of most plants is underground and serves as anchorage and for absorption of water and nutrients. The primary root originates in the embryo and is the first structure to emerge from the seed. Primary roots are usually positively geotropic i.e. they respond to gravity by growing downward.

Like the shoot, the root has an apical meristem. This root apical meristem (RAM) is covered with a protective root cap. This cap protects the root tip and the RAM as they grow down through the soil. The RAM produces cells toward the root body that differentiate into the vascular, ground, and dermal tissues of the root system. The RAM also adds new cells to the root cap.

While the vascular tissue in the shoot system is positioned around the circumference of the shoot, the vascular tissue is located in the central stele of the root. New lateral roots arise in the pericycle that surrounds the central column of xylem and phloem.

Many first order lateral roots do not respond to the forces of gravity and grow parallel with the ground surface as a means to finding an abundant source of water. Root hairs are formed by epidermal cells that extend outward from the root. This provides a huge amount of surface area to take up water. The root's region of elongation has no root hairs protruding from the epidermis because they would easily be torn or sloughed off as the root pushes down into the soil.

ALLORHIZIC VS. HOMORHIZIC RADICATION

Allorhizic refers to the fact that there are spatially separated and distinct shoot and root systems. The embryos of gymnosperms and dicotyledonous angiosperms are bipolar with a root or radicle pole directly opposite the shoot pole. The pteridophytes are homorhizic and exhibit no primary root pole and therefore lack the development of a separate root system. Their roots are initiated by the shoot system.

Monocotyledonous angiosperms are homorhizic as a result of the precocious death of the primary root that forms in the embryo. All subsequent roots that form on monocotyledons originate from the stem. These shoot-borne roots are called adventitious roots. Shoot borne roots may originate solely at nodal positions as seen in *Zea mays*, or they may originate internodally, as seen in *Hedera helix*, a dicot.

Tap Root and Fibrous Root Systems

In gymnosperms and dicots, the primary root becomes a taproot that produces lateral roots as it grows downward into the soil. In this type of root system, the older lateral roots are located near the soil surface.

Since the primary root in monocots is short-lived, some develop a system of adventitious roots with many lateral branches. This fibrous root system does not grow as deeply into the soil as a taproot system but the fibrous nature of the system enables them to stay well rooted in shallow soil and prevent erosion.

STRUCTURAL SPECIALIZATIONS OF ROOTS AND ROOT SYSTEMS

While the main functions of roots are to anchor the plant and to take up water and nutrients, some roots may serve other purposes in certain environments. Many epiphytes in tropical rainforests have highly modified roots that allow the plants to survive high in the rainforest canopy where water and nutrients are not so readily available.

Myrmecodia species have swollen hypocotyl regions covered with thorn roots. The swollen region contains chambers where ants live. The ants provide the plant with nutrients when the ants die and decay inside the plant, and the plant produces the thorn roots that prevent ant-eating predators from disturbing the ant's home.

Roots may also serve as storage areas for nutrients and/ or water. In radishes the hypocotyl region begins to expand as photosynthates, usually carbohydrates begin to accumulate. Eventually these reserves accumulate downward into the taproot. In *Pachypodium* this storage works in reverse: the root begins to swell with water and eventually water storage moves up into the hypocotyl region. Other examples of storage roots include beets, rutabagas, and carrots. Contractile roots are another type of root specialization. A contractile root can pull the shoot apical meristem below ground surface for overwintering. Some spring bulbs such as crocus undergo contraction against full soil resistance. As the root contracts, it pulls the bulb down into the soil and compacts the soil located under the bulb.

Some plants, like the dandelion, move without soil resistance. In this case, the root grows downward initially via cell elongation, then the cells expand laterally resulting in contraction of the root. There is no compaction of the soil and a channel is left behind the contracted root. The contracted portions of the roots on a dandelion plant make it virtually impossible to pull up a dandelion plant with the roots intact.

Stilt roots, such as those found on red mangrove (*Rhisophora*), allow these plants to live in a habitat that experiences frequent flooding. Since the substrate is often under the surface of the water, stem-borne stilt roots form a tripod-like support for the shoot system. Stilt roots are commonly found on plants with an obconate shoot form.

Some *Ficus* species have prop roots that originate from the lower surface of lateral branches. These prop roots support the lateral branches as they extend great distances from the main trunk of the tree.

Pneumatophores are specialized roots produced by some trees (e.g. *Taxodium*, bald cypress, *Avicennia*, black mangrove) adapted to life in swamps. *Pneumatophores* grow upward (i.e.

they are negatively geotropic) into the air above the anaerobic muck in which the rest of the root system is found. The surface of the woody pneumatophore is covered with lenticels that allow air to diffuse into the rest of the root system.

Aerial roots serve several purposes to epiphytic plants. Many orchids have aerial roots covered with velamen, multiple layers of dead epidermal cells that absorb water and nutrients directly form the air. The aerial roots of some epiphytic orchids have a cortex containing chloroplasts for photosynthesis. These aerial roots may be tendril-like and twine to hold the plant to its substrate. These roots show a thigmotropic response and twine around various objects.

SPECIALIZATIONS IN ROOT STRUCTURE

Symbioses are associations between individuals of two different species and are very widespread in nature. Parasitism is a form of symbiosis where one of the partners in the association is harmed, and mutualism is an association where both partners benefit. The association between root systems and microbes are often of the mutualistic type.

Nitrogen is the most abundant element in the earth's atmosphere (roughly 80%) but cannot be taken up directly by plants because plants lack the enzymes needed to break up the triple bond in N_2.

Nitrogen fixation is the conversion of this diatomic nitrogen to compounds that can be assimilated by plants and is accomplished only by certain bacteria. While some nitrogen-fixing bacteria live freely in the soil, others form mutualistic symbioses with plant roots. The bacteria in such symbioses receive photosynthates from the plant while the plant gets a usable form of nitrogen from the bacteria. These mutualistic bacteria live in nodules on the roots of 10 different angiosperm families including legumes.

The presence of these nitrogen-fixing nodules on crops minimizes the use of commercial fertilizers, and thus they are environmentally beneficial. A great deal of genetic research is being done in hopes of transferring the capacity to form nodules, or the capacity to fix nitrogen or both to other crops.

Certain plants release flavenoids into the soil that attract nitrogen-fixing bacteria. In response to the flavenoids the bacteria produce factors coded on nod genes. The nodule-forming plant responds to the nod factors by curling of root hairs and formation of infection threads by which the bacteria enter the host root. The bacteria take up residence in enlarged cells in the root cortex.

Root nodules may be indeterminate in nature, growing as long as the plant grows. This type of nodule is found in association with alfalfa and peas. Determinate nodules are found in association with soybeans.

Coralloid Roots of Cycads

Cycads produce specialized roots in response to mutualistic nitrogen-fixing cyanobacteria. Such roots are apogeotropic (uninfluenced in their growth by the force of gravity) and termed "coralloid" because they resemble dichotomously branched coral. Coralloid roots are occupied by the cyanobacterial (bluegreen algal) genera *Anabaena* or *Nostoc*. Since the algae require light for photosynthesis, the coralloid roots are commonly found near the soil surface.

Mycorrhizae

The root systems of many plants are associated with mycorrhizal fungi that are involved in the uptake of phosphorus and other essential nutrients including zinc, manganese, and copper. The fungal hyphae extend outward from the root, increasing the surface area for water and nutrient uptake. The hyphae that penetrate into the root's interior deposit the nutrients inside root. Mycorrhizae may be ectotrophic, forming a mantle of hyphae around the root periphery. The hyphae also penetrate the cortical intercellular spaces but not the protoplast of the host cells. In contrast, endotrophic mycorrhizae invade the cells of the host. Suberin in the Casparian strip of the root's endodermis prevents fungal hyphae from entering the root's vascular tissue.

Mycorrhizae can have an enormous impact on a plant's ability to grow. Without the essential nutrients provided by these root symbionts a plant will struggle to survive. Plants with mycorrhizal associations often have less extensive root systems than plants without such associations.

THE STEM

What are the features that distinguish a stem from a root? The stem is the ascending part of the axis bearing branches, leaves, flowers and fruits. It develops from the plumule of the embryo of a germinating seed. The stem bears nodes and internodes. The region of the stem where leaves are born are called nodes while internodes are the portions between two nodes. The stem bears buds, which may be terminal or axillary. Stem is generally green when young and later often become woody and dark brown. The main function of the stem is spreading out branches bearing leaves, flowers and fruits. It conducts water, minerals and photosynthates. Some stems perform the function of storage of food, support, protection and of vegetative propagation.

Modifications of Stem

The stem may not always be typically like what they are expected to be. They are modified to perform different functions. Underground stems of potato, ginger, turmeric, *zaminkand, Colocasia* are modified to store food in them. They also act as organs of perenation to tide over conditions unfavourable for growth. Stem tendrils which develop from axillary buds, are slender and spirally coiled and help plants to climb such as in gourds (cucumber, pumpkins, watermelon) and grapevines. Axillary buds of stems may also get modified into woody, straight and pointed thorns. Thorns are found in many plants such as *Citrus, Bougainvillea.* They protect plants from browsing animals. Some plants of arid regions modify their stems into flattened (*Opuntia*), or fleshy cylindrical *(Euphorbia)* structures.

They contain chlorophyll and carry out photosynthesis. Underground stems of some plants such as grass and strawberry, etc., spread to new niches and when older parts die new plants are formed. In plants like mint and jasmine a slender lateral branch arises from the base of the main axis and after growing aerially for some time arch downwards to touch the ground.

A lateral branch with short internodes and each node bearing a rosette of leaves and a tuft of roots is found in aquatic plants like *Pistia* and *Eichhornia*. In banana, pineapple and *Chrysanthemum*, the lateral branches originate from the basal and underground portion of the main stem, grow horizontally beneath the soil and then come out obliquely upward giving rise to leafy shoots.

STEM SUCCULENTS

Desert plants have evolved specializations to survive the lack of water in the arid regions. Annuals can complete their entire life cycle during the brief rainy periods in deserts. Some perennials store water within their stems and leaves. Succulents minimize water loss by closing their stomata during the day when the temperatures soar under the hot desert sun, opening them at night to take in CO_2. During prolonged periods of drought some plants may keep their stomata closed both day and night.

Some succulents have a reduced surface to volume ratio, developing a globose shape like the barrel cactus. While we generally think of the cactus family (Cactaceae) when discussing succulents, other plant families, including Euphorbiaceae and Asclepiadaceae, have species with shape characteristics adapted for water storage. Instead of forming a tree with a solid core of xylem, many stem succulents have a woody support network with large parenchymatous gaps where large volumes of water can be stored.

The root system of many succulents consists of a very deep taproot as well as "rain roots" that may only grow to a depth of 5 centimeters but extend laterally for several meters.

Many succulents lack foliage leaves or only produce leaves when water is available. Cacti can be distinguished from other succulents by the presence of areoles. An areole is thought to be a modified lateral bud, or axillary short shoot with lateral appendages homologous to leaves modified as spines. An areole is generally borne on the adaxial surface or apex of a podarium, a modified leaf base.

The podaria of some species have become fused into vertical ribs that cover the entire surface of the plant. The barrel cactus provides a good example of this type of ribbing. The ribs form an accordion shape that allows the plant to expand and increase the volume of water it can hold. The plant may lose up to 80% of the water it stores. As the stem loses water, it folds inward at the "valleys" between the podaria, and its volume decreases, minimizing the surface area exposed to direct sunlight.

THE LEAF

The leaf is a lateral, generally flattened structure borne on the stem. It develops at the node and bears a bud in its axil. The axillary bud later develops into a branch. Leaves originate from shoot apical meristems and are arranged in an acropetal order. They are the most important vegetative organs for photosynthesis. A typical leaf consists of three main parts: leaf base, petiole and lamina. The leaf is attached to the stem by the leaf base and may bear two lateral small leaf like structures called stipules.

In monocotyledons, the leaf base expands into a sheath covering the stem partially or wholly. In some leguminous plants the leafbase may become swollen, which is called the pulvinus. The petiole help hold the blade to light. Long thin flexible petioles allow leaf blades to flutter in wind, thereby cooling the leaf and bringing fresh air to leaf surface. The lamina or the leaf blade is the green expanded part of the leaf with veins and veinlets. There is, usually, a middle prominent vein, which is known as the midrib. Veins provide rigidity to the leaf blade and act as channels of transport for water, minerals and food materials. The shape, margin, apex, surface and extent of incision of lamina varies in different leaves.

Venation

The arrangement of veins and the veinlets in the lamina of leaf is termed as venation. When the veinlets form a network, the venation is termed as reticulate. When the veins run parallel to each other within a lamina, the venation is termed as parallel. Leaves of dicotyledonous plants generally possess reticulate venation, while parallel venation is the characteristic of most monocotyledons.

Types of Leaves

A leaf is said to be simple, when its lamina is entire or when incised, the incisions do not touch the midrib. When the incisions of the lamina reach up to the midrib breaking it into a number of leaflets, the leaf is called compound. A bud is present in the axil of petiole in both simple and compound leaves, but not in the axil of leaflets of the compound leaf. The compound leaves may be of two types.

In a pinnately compound leaf a number of leaflets are present on a common axis, the rachis, which represents the midrib of the leaf as in neem. In palmately compound leaves, the leaflets are attached at a common point, i.e., at the tip of petiole, as in silk cotton.

Phyllotaxy

Phyllotaxy is the pattern of arrangement of leaves on the stem or branch. This is usually of three types – alternate, opposite and whorled. In alternate type of phyllotaxy, a single leaf arises at each node in alternate manner, as in china rose, mustard and sun flower plants.

In opposite type, a pair of leaves arise at each node and lie opposite to each other as in *Calotropis* and guava plants. If more than two leaves arise at a node and form a whorl, it is called whorled, as in *Alstonia.*

Modifications of Leaves

Leaves are often modified to perform functions other than photosynthesis. They are converted into tendrils for climbing as in peas or into spines for defence as in cacti. The fleshy leaves of onion and garlic store food. In some plants such as Australian acacia, the leaves are small and short-lived. The petioles in these plants expand, become green and synthesise food. Leaves of certain insectivorous plants such as pitcher plant, venus-fly trap are also modified leaves.

VARIATIONS IN LEAF MORPHOLOGY

Leaves can be classified as either simple or compound. Simple leaves have a single blade. Although undivided into separate leaflets, the blade of a simple leaf may be lobed with deep sinuses.

Compound leaves are separated into two or more leaflets. Compound, or dissected leaves, are further classified as pinnate or palmate. Pinnately compound leaves are dissected very much like a feather, with leaflets located on both sides of a main axis called a rachis. Palmately compound leaves have leaflets that share a common insertion point at the end of the petiole. Both pinnately and palmately compound leaves can be divided again resulting in bipinnate or bipalmate leaf forms.

Differential growth within a developing leaf primordium determines the shape of the mature leaf. Initially, growth is continuous along the margin of a leaf primordium. After the leaf primordium becomes differentiated into a leaf blade and leaf base region, differential growth may occur along the primordial blade margin.

If growth is more or less continuous along the margin, the primordium will develop into a simple leaf. If growth is localized along the margin, the discrete areas of growth may develop into the separate leaflets of a compound leaf. All manner of intermediates can be found between leaves with a simple lamina and those with a dissected blade. Many simple leaf forms initiate what appear to be primordial pinnae or leaflets, but these develop into lobes or teeth because of continued growth between the initially separate areas of growth along the primordial margin.

Most palms have compound leaves that develop in a completely different way than described in the previous paragraph. Palm leaves start out as primoria with continuous growth along their margins. The palm leaf becomes separated into separate leaflets by the physical breakdown and tearing of tissues within the initially simple primordial leaf blade.

The breakage into separate leaflets occurs as the blade expands and grows out of the bud. Strips of unbroken tissue, called reins, may remain connecting the ends of the leaflets in both feather (pinnately compound) or fan (palmately compound) palm leaves.

Leaves of some plants, such as *Monstera* and *Philodendron,* in the family Araceae, have holes in them. These holes occur late during expansion due to apoptosis (programmed cell death) in those regions of the blade.

Some monocots (especially grasses and sedges) differ from typical dicots because their leaves retain basal intercalary meristems. This meristem, located near the base of the leaf, continues to produce new cells even in an otherwise fully mature leaf. As expansion occurs, the oldest cells are found near the tip of the leaf and the newest cells are located near the leaf base. This basal meristem explains why we can continually cut our lawns and the grass keeps growing back.

Some plants develop unifacial or equifacial leaves that apparently lack adaxial and abaxial surfaces. Instead of being flattened, the leaf blade is terete (cylindrical). In some plants such as *Musa* (banana), only the very tip of the leaf is unifacial and in others such as *Sanseveria cylindrica,* the entire leaf is unifacial.Peltate leaves have umbrella-shaped blades with a petiole that is attached in the middle of the abaxial surface.

LEAF SPECIALIZATION IN RELATION TO ECOLOGICAL ROLES

Most leaves function as photosynthetic organs, but some leaves serve specialized functions that enable a plant to survive in a particular environment. The photosynthetic capabilities of specialized leaves may be reduced or enhanced depending on the role they serve.

Cataphylls (bud scales) are modified leaves that cover and protect young vegetative or reproductive shoots. Cataphylls are often leathery or have a soft insulated lining that protects the bud over harsh winters. Many bud scales exude sticky substances that help keep the bud watertight and protect it from insect damage. It was once thought that bud scales were leaves whose growth was arrested at the end of the growing season. However, many bud scales have developmental patterns that are quite different from those of typical foliage leaves. The number and morphology of bud scales varies greatly between plants and these differences can be used for identification purposes.

Some plants, such as *Pisum*, have highly modified leaf tendrils. Leaves, particularly the upper leaf regions (petiole and lamina), may be modified into twining structures similar to the shoot tendrils of climbing plants. The hook tendrils found on *Cissus* vines provide an excellent example of this modification. These tendrils often mature into thick woody structures. Another good example of modified leaf tendrils is the adhesive disc tendril of *Parthenocissus*. The leaves of *Clematis* provide examples of petioles modified as tendrils. These petioles elongate greatly to twine around fences and trellises.

Window leaves allow some plants to survive in arid desert environments. Both *Haworthia* and *Lithops* are succulent plants that incorporate these transparent surfaces that direct light to their interiors. The stone plant, *Lithops*, has its succulent leaves mostly buried below ground and thus protected from desiccation. The leaf surfaces exposed to sunlight direct the light to leaf tissues below ground that contain chloroplasts.

Some epiphytic ferns, such as the staghorn fern, have leaves modified to clasp onto other plants or substrates and prevent them from falling. These epiphytic ferns are frequently found in the rainforest. Clasping leaves hold the plant to its substrate high in the canopy where it can successfully compete for sunlight and water.

LEAVES OF INSECTIVOROUS PLANTS

Wetlands like bogs are often highly acidic due to peat-forming mosses. Acidic conditions make nutrients such as nitrogen, potassium, phosphorous, calcium, and magnesium extremely difficult for most plants to take up through their roots. Plants have evolved to survive these conditions by trapping insects and digesting them to provide the nutrients needed for growth.

Many of these plants have structural modifications that include rosette shoots, highly modified leaves, and poorly developed root systems (because the leaves gather the nutrients). There are 3 categories of insectivorous plants based on the type of insect trap mechanism used by the plant: passive traps, active traps, and suction traps.

The pitcher plants such as *Nepenthes,Cephalotus,* and *Sarracenia* have passive traps. Insects simply fall into a pit formed by the tubular leaf blades of these plants. The insects may be lured to the traps by various colours, scents, and nectars. Once an insect has fallen into the water-filled trap its escape may be hindered by trichomes or slippery secretions.

Active traps such as the notorious Venus flytrap react to insects on their leaf surfaces by secreting digestive enzymes. *Pinguicula* and *Drosera,* the sundew, have hairs that bend inward and leaf margins that roll up to help trap the insects. *Dionaea,* the Venus flytrap, has leaf blades with trigger hairs that, when stimulated, will cause the hinged blade to quickly close over an insect.

The suction traps of *Utricularia,* the bladderwort, have leaves with small sacs, the bladders. An insect is trapped when it trips hairs around the opening of the bladder. This results in water being sucked into the bladder, carrying the insect with it. Once inside, the insect is digested.

FERN LEAF MORPHOLOGY

The fossil record indicates that ferns have lived on Earth for more than 350 million years. With thousands of species today, ferns are still highly diverse especially in the tropics.

Fern leaves, or fronds, undergo acropetal maturation with the oldest cells found near the base of the leaf. The tip of a fern leaf remains meristematic within a spiraled structure known as a fiddlehead or crozier. The uncoiling or maturation of the leaf is termed circinate vernation.

Some fern leaves may take up to five years to fully expand. Fronds of some ferns, such as *Nephrolepis,* exhibit intermittent apical growth so that its apical meristematic activity extends over several seasons. As the apical meristem generates new cells, they elongate the leaf causing the fiddlehead to be located further away from the petiole as the leaf matures. The climbing fern, *Lygodium*, is known to have fronds up to 45 feet in length. The lateral pinnae found on some ferns start out as croziers also and are known as subcroziers.

Most fern leaves are borne on a compact stem as a rosette or on a rhizome. The rosette types may either be ground-hugging, such as *Adiantum capillus-veneris*, or unbranched-tree-like, as in *Blechnum gibbum*, where the rosette is found at the end of a stem.

Some ferns (e. g. the walking fern *Asplenium*) have leaves that act as stolons: where the tip of the fern leaf touches the ground, a new plant is generated by activity of the leaf's apical cell.

Several ferns exhibit phyllogenous branching. *Hemionitus pinnatifida, Pellaea rotundifolia,* and other have leaves that give rise to branches.

Many ferns have leaves that perform both a reproductive (spore-producing) and a vegetative (photosynthetic) function. Other ferns have separate leaves specialized either to bear sporangia or to do photosynthesis. Sporangia are typically grouped together into structures called sori (sorus, singular). The location and form of sori is often taxonomically important.

Many ferns are epiphytes that clasp onto other plants high in the canopy or into rocky ledges where they are exposed to sunlight. Approximately one third of all tropical fern species exhibit this type of behaviour. The staghorn fern discussed in a previous section has clasping leaves that hold on to a tree or ledge acting as a substrate.

THE INFLORESCENCE

A flower is a modified shoot wherein the shoot apical meristem changes to floral meristem. Internodes do not elongate and the axis gets condensed. The apex produces different kinds of floral appendages laterally at successive nodes instead of leaves. When a shoot tip transforms into a flower, it is always solitary. The arrangement of flowers on the floral axis is termed as inflorescence.

Fig: Racemose Inflorescence

Depending on whether the apex gets converted into a flower or continues to grow, two major types of inflorescences are defined – racemose and cymose. In racemose type of inflorescences the main axis continues to grow, the flowers are borne laterally in an acropetal succession.

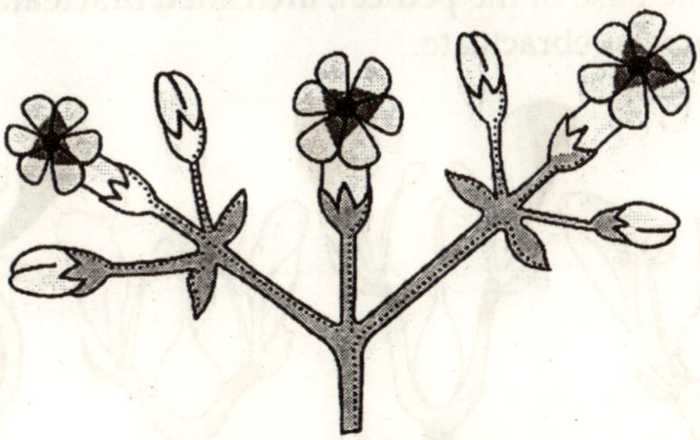

Fig: Cymose Inflorescence

In cymose type of inflorescence the main axis terminates in a flower, hence is limited in growth.The flowers are borne in a basipetal order.

THE FLOWER

The flower is the reproductive unit in the angiosperms. It is meant for sexual reproduction. A typical flower has four different kinds of whorls arranged successively on the swollen end of the stalk or pedicel, called thalamus or receptacle. These are calyx, corolla, androecium and gynoecium. Calyx and corolla are accessory organs, while androecium and gynoecium are reproductive organs. In some flowers like lily, the calyx and corolla are not distinct and are termed as perianth. When a flower has both androecium and gynoecium, it is bisexual.

A flower having either only stamens or only carpels is unisexual. In symmetry, the flower may be actinomorphic (radial symmetry) or zygomorphic (bilateral symmetry). When a flower can be divided into two equal radial halves in any radial plane passing through the centre, it is said to be actinomorphic, e.g., mustard, *datura*, chilli. When it can be divided into two similar halves only in one particular vertical plane, it is zygomorphic, e.g., pea, gulmohur, bean, *Cassia*.

A flower is asymmetric (irregular) if it cannot be divided into two similar halves by any vertical plane passing through the centre, as in canna. A flower may be trimerous, tetramerous or pentamerous when the floral appendages are in multiple of 3, 4 or 5, respectively. Flowers with bracts, reduced leaf found at the base of the pedicel, are called bracteate and those without bracts, ebracteate.

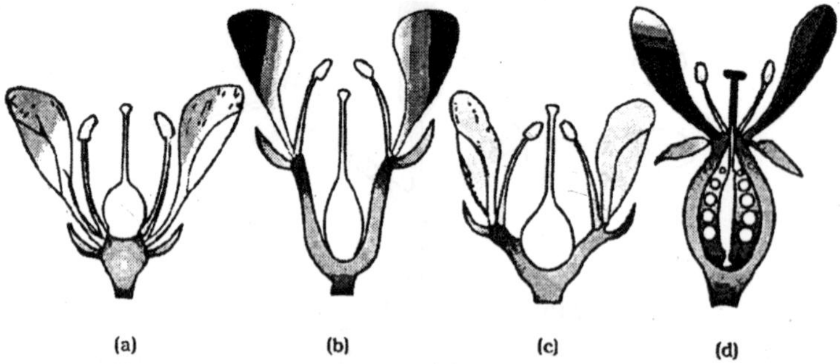

(a) (b) (c) (d)

Fig: Position of floral parts on thalamus:
(a) Hypogynous (b) and (c) Perigynous (d) Epygynous

Based on the position of calyx, corolla and androecium in respect of the ovary on thalamus, the flowers are described as hypogynous perigynous and epigynous. In the hypogynous flower the gynoecium occupies the highest position while the other parts are situated below it. The ovary in such flowers is said to be superior, e.g., mustard, china rose and brinjal.

If gynoecium is situated in the centre and other parts of the flower are located on the rim of the thalamus almost at the same level, it is called perigynous. The ovary here is said to be half inferior, e.g., plum, rose, peach. In epigynous flowers, the margin of thalamus grows upward enclosing the ovary completely and getting fused with it, the other parts of flower arise above the ovary. Hence, the ovary is said to be inferior as in flowers of guava and cucumber, and the ray florets of sunflower.

Parts of a Flower

Each flower normally has four floral whorls, viz., calyx, corolla, androecium and gynoecium

Calyx

The calyx is the outermost whorl of the flower and the members are called sepals. Generally, sepals are green, leaf like and protect the flower in the bud stage. The calyx may be gamosepalous (sepals united) or polysepalous (sepals free).

Corolla

Corolla is composed of petals. Petals are usually brightly coloured to attract insects for pollination.

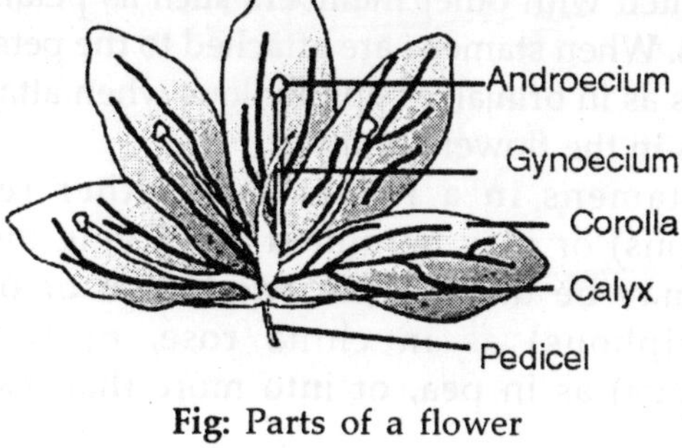

Fig: Parts of a flower

Like calyx, corolla may be also free (gamopetalous) or united (polypetalous). The shape and colour of corolla vary greatly in plants. Corolla may be tubular, bell-shaped, funnel-shaped or wheel-shaped.

Aestivation

The mode of arrangement of sepals or petals in floral bud with respect to the other members of the same whorl is known as aestivation. The main types of aestivation are valvate, twisted, imbricate and vexillary. When sepals or petals in a whorl just touch one another at the margin, without overlapping, as in *Calotropis*, it is said to be valvate.

If one margin of the appendage overlaps that of the next one and so on as in china rose, lady's finger and cotton, it is called twisted. If the margins of sepals or petals overlap one another but not in any particular direction as in *Cassia* and gulmohur, the aestivation is called imbricate. In pea and bean flowers, there are five petals, the largest (standard) overlaps the two lateral petals (wings) which in turn overlap the two smallest anterior petals (keel); this type of aestivation is known as vexillary or papilionaceous.

Androecium

Androecium is composed of stamens. Each stamen which represents the male reproductive organ consists of a stalk or a filament and an anther. Each anther is usually bilobed and each lobe has two chambers, the pollen-sacs. The pollen grains are produced in pollen-sacs.

A sterile stamen is called staminode. Stamens of flower may be united with other members such as petals or among themselves. When stamens are attached to the petals, they are epipetalous as in brinjal, or epiphyllous when attached to the perianth as in the flowers of lily.

The stamens in a flower may either remain free (polyandrous) or may be united in varying degrees. The stamens may be united into one bunch or one bundle (monoadelphous) as in china rose, or two bundles (diadelphous) as in pea, or into more than two bundles (polyadelphous) as in citrus. There may be a variation in the length of filaments within a flower, as in *Salvia* and mustard.

Gynoecium

Gynoecium is the female reproductive part of the flower and is made up of one or more carpels. A carpel consists of three parts namely stigma, style and ovary. Ovary is the enlarged basal part, on which lies the elongated tube, the style. The style connects the ovary to the stigma. The stigma is usually at the tip of the style and is the receptive surface for pollen grains. Each ovary bears one or more ovules attached to a flattened, cushion-like placenta. When more than one carpel is present, they may be free (as in lotus and rose) and are called apocarpous.

They are termed syncarpous when carpels are fused, as in mustard and tomato. After fertilisation, the ovules develop into seeds and the ovary matures into a fruit. Placentation: The arrangement of ovules within the ovary is known as placentation. The placentation are of different types namely, marginal, axile, parietal, basal, central and free central. In marginal placentation the placenta forms a ridge along the ventral suture of the ovary and the ovules are borne on this ridge forming two rows, as in pea. When the placenta is axial and the ovules are attached to it in a multilocular ovary, the placentaion is said to be axile, as in china rose, tomato and lemon. In parietal placentation, the ovules develop on the inner wall of the ovary or on peripheral part. Ovary is one-chambered but it becomes twochambered due to the formation of the false septum, e.g., mustard and *Argemone*.

When the ovules are borne on central axis and septa are absent, as in *Dianthus* and *Primrose* the placentation is called free central. In basal placentation, the placenta develops at the base of ovary and a single ovule is attached to it, as in sunflower, marigold.

THE FRUIT

The fruit is a characteristic feature of the flowering plants. It is a mature or ripened ovary, developed after fertilisation. If a fruit is formed without fertilisation of the ovary, it is called a parthenocarpic fruit. Generally, the fruit consists of a wall or pericarp and seeds. The pericarp may be dry or fleshy. When pericarp is thick and fleshy, it is differentiated into the outer epicarp, the middle mesocarp and the inner endocarp. In mango and coconut, the fruit is known as a drupe.

They develop from monocarpellary superior ovaries and are one seeded. In mango the pericarp is well differentiated into an outer thin epicarp, a middle fleshy edible mesocarp and an inner stony hard endocarp. In coconut which is also a drupe, the mesocarp is fibrous.

THE SEED

The ovules after fertilisation, develop into seeds. A seed is made up of a seed coat and an embryo. The embryo is made up of a radicle, an embryonal axis and one (as in wheat, maize) or two cotyledons (as in gram and pea).

Structure of a Dicotyledonous Seed

The outermost covering of a seed is the seed coat. The seed coat has two layers, the outer testa and the inner tegmen. The hilum is a scar on the seed coat through which the developing seeds were attached to the fruit. Above the hilum is a small pore called the micropyle. Within the seed coat is the embryo, consisting of an embryonal axis and two cotyledons.

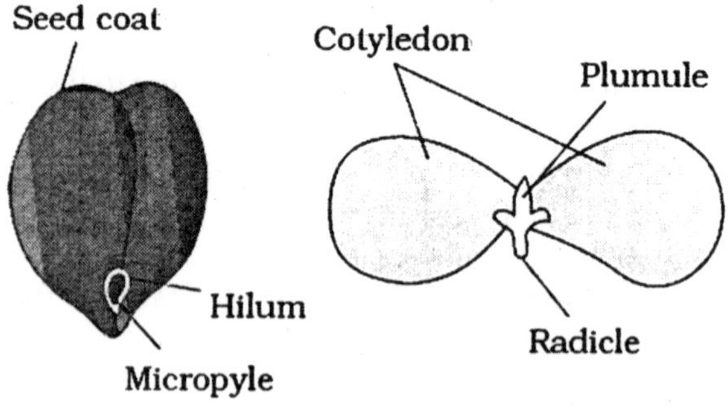

Fig: Structure of a Dycotyledonous Seed

The cotyledons are often fleshy and full of reserve food materials. At the two ends of the embryonal axis are present the radicle and the plumule. In some seeds such as castor the endosperm formed as a result of double fertilisation, is a food storing tissue. In plants such as bean, gram and pea, the endosperm is not present in mature seeds and such seeds are called non-endospermous.

Structure of Monocotyledonous Seed

Generally, monocotyledonous seeds are endospermic but some as in orchids are non-endospermic. In the seeds of cereals such as maize the seed coat is membranous and generally fused with the fruit wall.

The endosperm is bulky and stores food. The outer covering of endosperm separates the embryo by a proteinous layer called aleurone layer.

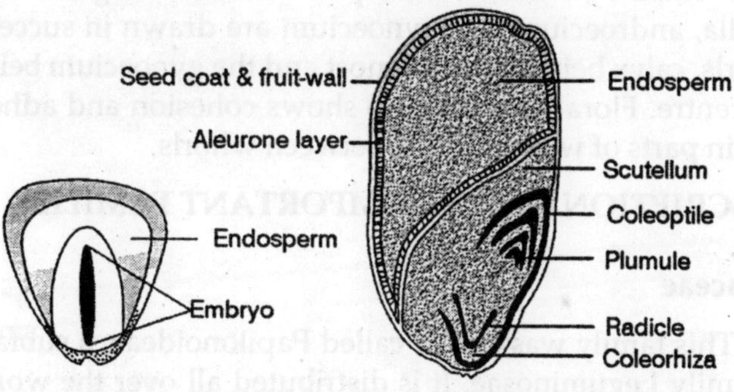

Fig: Structure of a Monocotyledonous

The embryo is small and situated in a groove at one end of the endosperm. It consists of one large and shield shaped cotyledon known as scutellum and a short axis with a plumule and a radicle. The plumule and radicle are enclosed in sheaths which are called coleoptile and coleorhiza respectively.

SEMI-TECHNICAL DESCRIPTION OF A TYPICALFLOWERING PLANT

Various morphological features are used to describe a flowering plant. The description has to be brief, in a simple and scientific language and presented in a proper sequence. The plant is described beginning with its habit, vegetative characters – roots, stem and leaves and then floral characters inflorescence and flower parts.

After describing various parts of plant, a floral diagram and a floral formula are presented. The floral formula is represented by some symbols. In the floral formula, Br stands for bracteate K stands for calyx, C for corolla, P for perianth, A for androecium and G for Gynoecium, G for superior ovary and G for inferior ovary, for male, for female, for bisexual plants,for actinomorphic and for zygomorphic nature of flower. Fusion is indicated by enclosing the figure within bracket and adhesion by a line drawn above the symbols of the floral parts. A floral diagram provides information about the number of parts of a flower, their arrangement and the relation they have with one another. The position of the mother axis with respect to the flower is represented by a dot on the top of the floral diagram. Calyx, corolla, androecium and gynoecium are drawn in successive whorls, calyx being the outermost and the gynoecium being in the centre. Floral formula also shows cohesion and adhesion within parts of whorls and in between whorls.

DESCRIPTION OF SOME IMPORTANT FAMILIES

Fabaceae

This family was earlier called Papilonoideae, a subfamily of family Leguminosae. It is distributed all over the world.

Vegetative Characters

Trees, shrubs, herbs; root with root nodules

Stem: erect or climber

Leaves: alternate, pinnately compound or simple; leaf base, pulvinate; stipulate; venation reticulate. Floral characters

Inflorescence: racemose

Flower: bisexual, zygomorphic

Calyx: sepals five, gamosepalous; imbricate aestivation

Corolla: petals five, polypetalous, papilionaceous, consisting of a posterior standard, two lateral wings, two anterior ones forming a keel (enclosing stamens and pistil), vexillary aestivation

Androecium: ten, diadelphous, anther dithecous

Gynoecium: ovary superior, mono carpellary, unilocular with many ovules, style single

Fruit: legume; seed: one to many, non-endospermic

Economic importance

Many plants belonging to the family are sources of pulses (gram, *arhar,sem, moong*, soyabean; edible oil (soyabean, groundnut); dye (indigofera); fibres (sunhemp); fodder *(Sesbania, Trifolium)*, ornamentals (lupin, sweet pea); medicine (*muliathi*).

Solanaceae

It is a large family, commonly called as the 'potato family'. It is widely distributed in tropics, subtropics and even temperate zones.

Vegetative Characters

Plants mostly, herbs, shrubs and small trees Stem: herbaceous rarely woody, aerial; erect, cylindrical, branched, solid or hollow, hairy or glabrous, underground stem in potato (*Solanum tuberosum*)

Leaves: alternate, simple, rarely pinnately compound, exstipulate; venation reticulate

Floral Characters

Inflorescence: Solitary, axillary or cymose as in *Solanum*

Flower: bisexual, actinomorphic

Calyx: sepals five, united, persistent, valvate aestivation

Corolla: petals five, united; valvate aestivation

Androecium: stamens five, epipetalous

Gynoecium: bicarpellary, syncarpous; ovary superior, bilocular, placenta swollen with many ovules

Fruits: berry or capsule

Seeds: many, endospermous

Economic Importance

Many plants belonging to this family are source of food (tomato, brinjal, potato), spice (chilli); medicine (belladonna, *ashwagandha*); fumigatory (tobacco); ornamentals (petunia).

Lilaceae

Commonly called the 'Lily family' is a characteristic representative of monocotyledonous plants. It is distributed world wide.

Vegetative characters: Perennial herbs with underground bulbs/corms/ rhizomes

Leaves mostly basal, alternate, linear, exstipulate with parallel venation

Floral characters Inflorescence: solitary / cymose; often umbellate clusters

Flower: bisexual; actinomorphic

Perianth tepal six (3+3), often united into tube; valvate aestivation

Androcium: stamen six, (3+3)

Gynoecium: tricarpellary, syncarpous, ovary superior, trilocular with many ovules; axile placentation

Fruit: capsule, rarely berry

Seed: endospermous

Economic Importance

Many plants belonging to this family are good ornamentals (tulip, *Gloriosa*), source of medicine (*Aloe*), vegetables (*Asparagus*), and colchicine (*Colchicum autumnale*).

AQUATIC PLANTS

A major evolutionary advance was the adaptation of early members of the plant kingdom to a terrestrial environment. Although most plants are restricted to life on land, some have become secondarily adapted to the habitat of their green algal ancestors. Light is a limiting factor in aquatic environments, and light availability becomes more of a problem with greater depth. The availability of oxygen and nutrients may also be limiting underwater. Water turbulence can be especially hazardous to aquatic plants. Specialized morphological features enable plants to overcome some of these difficulties of living in water.

Some aquatic plants, such as *Nymphaea*, have floating leaves while their root systems are anchored to the bottom of the lake or river and the shoot remains just above the substrate. The shoot apical meristem produces leaves with elongated petioles that allow the leaf blade to float on the surface of the water.

The petiole contains aerenchyma, a spongy tissue with large air spaces found between the cells, that helps the leaf float. As water level rises and falls, the flexible petiole allows the blade to remain at the water's surface. The leaves are often peltate in shape. When leaves become crowded together some shading of neighbouring leaves may occur. Victoria regina, the royal water lily, has huge peltate leaves with upturned leaf margins that prevent the leaves from overlapping.

Completely submerged aquatic plants can only grow in water bodies that are shallow enough to permit light to penetrate to them. Many have rosette shoots and highly dissected leaf blades. *Aponogeton fenestrale* leaf blades have vascular tissue surrounded by a single layer of chlorenchyma. The spaces between the vasculature result from apoptosis (programmed cell death). The spaces allow highly turbulent water to pass through the leaf blade without shredding it.

Free-floating aquatic plants are not rooted in a substrate but exist entirely on the water's surface. Some species have root-like structures that facilitate nutrient absorption and others have reduced root systems that act as a keel to keep the plant upright.

Eichornia crassipes, the water hyacinth, has leaves with aerenchyma filled bulbous petioles that keep them afloat. *Pistia* has leaves covered with hairs that "trap" air bubbles to ensure the plant stays above water. Many free-floating aquatic species are extremely small. *Wolffia* and *Spirodela* are examples of these small aquatic plants that tend to form colonies on the water's surface.

Free-floating ferns exhibit the same kinds of structural adaptations to an aquatic environment as the angiosperms mentioned above. *Azolla filiculoides* has a single root that acts like a keel. The leaves of *Azolla* are inhabited by a nitrogen-fixing bluegreen alga. These water ferns are grown on rice paddies and provide a source of nitrogen to the rice plants when the ferns die. *Salvinia* is a free-floating fern with three leaves per node: two trichome-covered leaves remain above water and one submerged highly dissected leave that looks like a root but bears sporangia.

Amphibious plants can live in water or on land. Some amphibious plants (e.g. *Ranunculus aquatilis*) are heterophyllous. The leaves that develop under water are highly dissected like the leaves of many aquatic plants. The leaves that develop in contact with the air are entire.

EPIPHYTISM IN THE FERNS AND ANGIOSPERMS

Epiphytes are plants that grow upon other plants, but unlike parasitic plants do not take food or water from their hosts. Organic material on the host's surface may provide nutrients. There are holoepiphytes, species that spend their entire lives as epiphytes, and hemiepiphytes, species that may spend part of their lives on a host and later become terrestrial.

The canopy of the tropical rain forest is a habitat where there is an abundance of epiphytes. While it is dark at the rain forest floor, in the canopy more light is available for photosynthesis. However, since epiphytes are not rooted in the soil they have difficulty getting water and inorganic nutrients.

Some epiphytes have structural adaptations that increase water uptake. The aerial roots of some epiphytes can take up moisture from the humid air. The velamen covering the aerial roots of some orchids may function in the absorption of water and also reduce water loss. Some bromeliads have rosette shoot systems that form "tanks" that can catch and hold large quantities of rainfall. Some epiphytes are succulent and store water in their stems.

Tillandsia usneoides, Spanish moss, is a common epiphyte found draped from the branches of trees in humid habitats such as those found around the Gulf of Mexico. Water and nutrients may be captured from water running down the stem of the host. More commonly, water and nutrients are absorbed from moisture in the air via specialized hygroscopic trichomes.

Ants are critical to many epiphytes. The ant plant, *Myrmecodia*, has a swollen shoot containing chambers that serve as a home for ants. The ants bring debris into the chambers and provide the plant with nutrients. *Dischidia rafflesiana* is another epiphyte that provides a home to ants.

This vine develops inflated "pocket" leaves that are inhabited by ants. These pockets also collect rainfall. Roots form at the node above the pocket leaf and grow into the pocket and take up the water from the rain and the nutrients left by the ants.

Epiphytic plants hold onto their hosts by specialized attachment roots that can enter crevices on the bark of their host or twine around branches. Many aerial roots are photosynthetic. Some epiphytic orchids lack foliage leaves and depend solely on their roots for photosynthesis. These orchids usually have minimal shoots that produce flowers for reproduction. Some epiphytic ferns, like the staghorn fern mentioned in a previous section, develop clasping or mantle leaves that keep the plant firmly attached to the host.

COMPARATIVE MORPHOLOGY OF PARASITIC FLOWERING PLANTS

Parasitic plants are those that live on or in another plant of a different species and derive their nutrients from their host. This type of association is beneficial to the parasite but harmful or lethal to the host. Parasitic plants are found worldwide, from tropical to sub-arctic regions.

Parasitic plants are categorized as root parasites or shoot parasites, depending on the portion of the host plant the parasite invades. Root and shoot parasites can be further subdivided into holoparasites, obligate and incapable of independent survival, or hemiparasites, able to survive on their own but with less vigour. Some parasitic plants may have extensive ectophytic (external to the host) systems and/or reduced endophytic (internal to the host) systems, while others may be largely endophytic in nature.

Root parasites may live underground for most of their life cycle. It is often not known that these parasites are present on the roots until the parasite becomes reproductive and sends floral shoots above ground. A parasitic plant may respond to signals given off by the roots of the host plant which normally attract fungi that are symbiotic with the host's roots. Instead of a helpful symbiont responding, a harmful parasite like *Striga* or *Orobanche* is attracted to the signal. The parasite invades by forming a haustorium (usually considered to be a modified root of the parasite) that grows into the host root and taps into its vascular cylinder. The haustorium can then take up water and/or nutrients from the host's vascular system.

Shoot parasites sometimes have larger ectophytic systems than root parasites. Many times their shoot systems are achlorophyllous and depend on their hosts for needed photosynthates. *Cuscuta*, the genus of dodder, is an achlorophyllous shoot parasite that has a vining habit. Once the dodder plant makes contact with and invades a host, the portion of the dodder plant rooted in the soil withers and dies. Dodder can easily destroy some crop plants, and if let go to seed, it can be very difficult to eradicate from a cultivated field.

The mistletoes, (the genera *Viscium, Arceuthobium,* and *Phoradendron*) invade tree branches. The mistletoes exhibit varying degrees of parasitism and reduction of the ectophytic system in favour of extensive development of the internal endophytic system. *Viscium album* has green photosynthetic shoots that are quite extensive and probably are nutritionally capable on their own. *Phoradendron* species show a more extensive development of the endophytic system. *Arceuthobium*, the dwarf mistletoe, has no ectophytic system and produces a very widely ramifying endophytic, mycelial-like plant body.

VARIATIONS IN THE POSITIONS OF REPRODUCTIVE SHOOTS IN VASCULAR PLANTS

Most treatments of reproductive morphology tend to isolate the reproductive organs from the remainder of the plant. Particular attention should be paid to the vegetative part of the shoot system and what, if any, transitions occur between the vegetative and reproductive phases of an individual shoot axis. This survey of reproductive morphology will focus on the cast of characters introduced in section one. The reproductive structures of angiosperms, the flowering plants.

Pteridophytes

Psilotum nudum has aerial shoots with bases that bear no appendages. Reduced leaves called enations develop further along the shoot. As the shoot becomes reproductive, the stem thickens and the enations are considered to be sporophylls (sporangium bearing leaves). The sporophylls, unlike the vegetative enations, are bifurcated. Each sporophyll subtends a synangium formed from three fused sporangia. At maturation the sporangia break open releasing a multitude of tiny spores. Following a reproductive phase, the shoots of the whisk fern may return to a vegetative state.

Selaginella develops sporophyll-bearing strobili at the tips of some or every stem, depending on the species and when they begin their reproductive phase. These cones can be hard to distinguish from the tips of vegetative shoots. *Selaginella* is heterosporous, and thus produces two kinds of spores. Each sporophyll bears a single sporangium on its adaxial surface. Megasporangia are borne by megasporophylls and microsporangia are borne by microsporophylls. Both kinds of sporangia occur in the same strobilus or cone.

Lycopodium is homosporous (one kind of spore). It produces sporangia on the upper surface of sporophylls. The sporophylls may or may not form an obvious strobilus.

Equisetum is also homosporous. Sporangia are borne along the margins of sporangiophores (sporangia-bearing branches) that form on strobili. A strobilus subtended by a peduncle, or stalk, is borne on the tip of either a lateral or terminal shoot. The peltate sporangiophores protect the spore-producing sporangia. Each sporangium has a line of dehiscence located in the internal position. Because the sporangium splits open internally, the spores are not dispersed very far from the mother plant.

Some species of *Equisetum* produce separate fertile (reproductive) shoots and vegetative shoots. The fertile shoots are achlorophyllous and depend on the vegetative shoots to provide them with photosynthates. *Equisetum sylvaticum* is one such species. In other species, such as *Equisetum hyemale*, the vegetative shoot becomes reproductive and produces an apical strobilus. *Equisetum myriochaetum* produces reproductive strobili on lateral shoots as well as the apical shoot.

There are marked convergences in form and behaviour between individual fern leaves and whole shoot segments in seed plants. The spores of ferns are produced in sporangia in clusters called sori which generally occur on the abaxial surface of the fronds. The sori may consist of small circular dots, stripes, patches, or broader areas covering nearly the entire abaxial leaf surface.

A unique property of most fern leaves is that they can be photosynthetic and reproductive organs simultaneously. There are some taxa that show either a marked separation of the reproductive and assimilative functions into separate fronds (dimorphism) or the bearing of sporangia on separate pinnae (pinna dimorphism). In such spatial separation of sporangia on the frond they represent analogies with the arrangement of seed plant reproductive structures on whole shoots.

Gymnosperms

The gymnosperms are plants that produce seeds not enclosed in an ovary. The seeds are usually borne in (strobili) cones. Megasporangiate cones produce the megaspores (female equivalent) and microsporangiate cones produce the microspores (male equivalent). While most conifers are monoecious, *Ginkgo biloba* and the cycads are dioecious. The strobili of cycads consist of reduced leaves with attached sporangia that are tightly clustered into cone-like structures near the apex of the plant.

The ovules of *Ginkgo* trees are produced in pairs at the end of short shoots. The ovules are protected within a fleshy seed coat and are not fertilized until they have fallen from the tree.

The microsporangia and megasporangia of most conifers are borne on the same tree. The microsporangiate cones are considered to be simple cones: the main axis bears microsporophylls directly. Male cones are usually produced on the lower branches of the tree. The megasporangiate strobili are complex cones: the main axis bears seed scale complexes homologous to lateral shoots subtended by bracts. The female strobili are usually produced in the upper branches, although some pines bear both male and female cones on the same branch.

At the time of pollination, the scales of megasporangiate cones are spread apart but grow together once pollination has occurred. It may take up to 3 years from pollination before the seeds are released from the cone. At maturity, the cone scales separate and the winged seeds are dispersed by the wind.

FLORAL MORPHOLOGY IN ANGIOSPERMS

Plants go through morphological changes throughout their lives. Some of these changes may be related to a transition to a reproductive phase. Many plants go through a juvenile phase when they are not capable of flowering under any conditions. Quite commonly, as in the case of *Acacia,* leaf morphology changes as the plant matures. Such a change is termed heteroblasty. Anatomical changes also may occur in leaves.

Juvenile leaves of *Zea mays* (corn) have a waxy coating that is not found on leaves at nodes in the reproductive zone. Phyllotactic changes may also occur during the transition to flowering. In sunflowers, the individual flowers are generated very rapidly, increasing the number of spirals. Many plants with rosette shoots during their vegetative phase undergo internodal elongation or bolting when stimulated to flower.

Once a plant reaches the reproductive phase, the shoot apical meristem is "reprogrammed" to produce flowers instead of leaves. When the shoot apical meristem is converted to a flower part (a carpel), the plant is determinate and it sacrifices itself for the production of seeds, its progeny. When the shoot apical meristem is not converted to a flower part, the meristem may produce flowers, yet the shoot apical meristem remains vegetative. Following a reproductive phase, the plant may return to a strictly vegetative state and the shoot apical meristem produces only leaves.

A flower consists of four whorls of leaf homologs: the sepals, the petals, the stamens, and the carpels. Flowers having all four floral whorls are considered complete and perfect. Flowers having either stamen or carpels, but not both, are termed imperfect. Evolutionary advantage is the driving force behind the vast variety of floral morphology.

Floral symmetry is one of many features analysed when examining floral morphology. Longitudinal symmetry (looking at a flower side on) allows one to determine where internodal development takes place, if there are variations in internodal thickening along the length of the floral shoot, and ovary position in relation to axis extension. (e.g. Does the flower have a superior ovary, an inferior ovary, or an intermediate position?)

Looking downward at the flower enables one to determine the flower's transectional symmetry. The possible symmetries include radial, bilateral, and dorsiventral. Phyllotactic patterns of the floral appendages may be similar throughout or may differ along the length of the floral shoot.

Fusion of floral parts is common throughout the angiosperms. Fusion between members of the same floral whorl, petal within the corolla, is termed connation. Fusion between members of different whorls (e.g. stamens fused with petals) is termed adnation. These conditions are evolutionarily advantageous because they function as pollination mechanisms.

Colour and texture variations occur on floral parts and are usually means of attracting pollinators. The colours of flowers may appear different to insects than to humans because they are able to perceive colours differently. Many flowers when viewed under ultraviolet light provide insects with a "bulls-eye," directing them to the reproductive structures. Of course, the insects are not concerned with pollinating the flowers but are interested in the pollen or nectar they use as food.

Night pollinators like bats and moths are attracted to white flowers, not by their colour, but by the strong scents they exude. The velvety appearance of some petals is caused by horizontal striations of cells making up the epidermis. A matte-like finish is the result of vertical striations of these cells. Conical epidermal cells, such as those found on the petals of the common marigold, reflect the wavelengths of light that attract pollinators.

Many plants are capable of self-pollination. Their reproductive structures mature simultaneously to insure that the stigmatic surface is prepared to accept a mature pollen grain. Some flowers may remain closed until fertilization occurs. There are many plants that prevent self-pollination. Their carpels and stamen may mature at different times. This type of reproductive selectivity assures a constantly changing gene pool and promotes evolution.

Flowering plants exhibit enormous variation in shape, size, structure, mode of nutrition, life span, habit and habitat. They have well developed root and shoot systems. Root system is either tap root or fibrous. Generally, dicotyledonous plants have tap roots while monocotyledonous plants have fibrous roots. The roots in some plants get modified for storage of food, mechanical support and respiration. The shoot system is differentiated into stem, leaves, flowers and fruits. The morphological features of stems like the presence of nodes and internodes, multicellular hair and positively phototropic nature help to differentiate the stems from roots. Stems also get modified to perform diverse functions such as storage of food, vegetative propagation and protection under different conditions.

Leaf is a lateral outgrowth of stem developed exoge-neously at the node. These are green in colour to perform the function of photosynthesis. Leaves exhibit marked variations in their shape, size, margin, apex and extent of incisions of leaf blade (lamina). Like other parts of plants, the leaves also get modified into other structures such as tendrils, spines for climbing and protection respectively. The flower is a modified shoot, meant for sexual reproduction. The flowers are arranged in different types of inflorescences. They exhibit enormous variation in structure, symmetry, position of ovary in relation to other parts, arrangement of petals, sepals, ovules etc. After fertilisation, the ovary is converted into fruits and ovules into seeds. Seeds either may be monocotyledonous or dicotyledonous.

They vary in shape, size and period of viability. The floral characteristics form the basis of classification and identification of flowering plants. Through semi-technical descriptions of families. Hence, a flowering plant is described in a definite sequence by using scientific terms. The floral features are represented in the summarised form as floral diagrams and floral formula.

Chapter 3

Pollination and Fertilization

Pollination is the transfer of pollen to the female organs of seed plants. In flowering plants (angiosperms, or "covered seeds"), immature seeds (ovules) are located within carpels. In contrast, nonflowering seed plants have uncovered ovules to which the pollen is transferred, making these "naked seeds" (gymnosperms).Gymnosperms have simpler pollination as all transmit their pollen by wind.

In contrast, angiosperm have a wealth of pollination methods involving many different agents to transfer pollen, including insects (entomophily), birds (ornithophily), bats (chirophily), wind (anemophily), and water (hydrophily). Attraction of animals usually occurs through a conspicuous floral display, with colour and scent playing important roles. Yellow and blue flowers tend to attract bees, red flowers attract hummingbirds, pink flowers attract butterflies, and white flowers that stand out at night often attract moths and bats. Animal-based pollination is efficient and usually associated with a food reward to assure a continuing relationship. Frequently, this is a simple symbiotic relationship; however, other plants seem to practice deceit.

Trap pollinators may hold insects hostage until a flower is pollinated successfully, when they are released. A most unusual mechanism are orchids that seemingly imitate the form and scent of female wasps, attracting amorous male wasps to mount the flower, inadvertently pollinating it. Aquatic plants frequently have filamentous pollen, which is more easily captured in water, and may release male flowers that float freely to their attached female counterparts.

When pollen is deposited on the stigma (in angiosperms) or the ovule (in gymnosperms), it germinates, forming a slender pollen tube through a weakened area of the pollen wall. The pollen tube elongates through "tip extension," penetrating between cells of the host parent. Within the pollen tube, two nonmotile sperm cells are ultimately formed and are conveyed through the tube, keeping pace with tip growth. The pollen tube uses chemotropic signals to determine the final pathway to the egg cell, deep within the ovule. In angiosperms, pollen tubes penetrate the stigma, style, and ovary until they are amid the ovules.

In gymnosperms, pollen germinates directly on the ovule. Pollen tubes enter ovules through a tiny pore called the micropyle and then elongate into the female gametophyte (called the embryo sac in angiosperms). In gymnosperms, the pollen tube directly penetrates the egg cell, but in angiosperms, there are sterile cells in the embryo sac, called synergids, that initially receive the sperm.

At this point, one sperm cell is discharged from the pollen tube and fuses, with the egg cell to form the zygote (the immediate fusion product) and subsequent embryo, which will become the offspring plant. In angiosperms, the second sperm fuses with the central cell to form a nutritive endosperm during double fertilization. The endosperm is needed for successful embryo development.

During fertilization the male and female gametes: (1) contact one another, (2) adhere, (3) cells fuse, and finally (4) nuclei fuse. The act of fertilization triggers embryo development in all plants and endosperm development in angiosperms.

FLOWERS

Flowers are the typically showy reproductive organs of angiosperms (flowering plants). Their diverse blooms generate countless horticultural products; functionally flowers are essential for sexual reproduction. Flowers may occur individually on the plant or together with other flowers forming inflorescences. Their flowers and inflorescences are:

(1) highly adapted to the plant's ecology and pollination strategy; (2) highly variable among the flowering plants; (3) critically important to plant identification; and (4) subject to their own rich and specialized terminology.

Morphologically and evolutionarily, the flower is regarded as a terminal shoot with variably leaflike lateral appendages. The stem of the flower, known as a pedicel, is connected to the other floral parts via the receptacle. From the outside rim of the receptacle, the first appendages are typically green, leaflike sepals, collectively called the calyx. The next whorl is composed of variously coloured and often complex petals, collectively called the corolla.

The androecium forms the next whorl, formed of stamens, each made of a filament and anther. Anther sacs are filled with pollen at maturity. Anthers release the pollen timed to match the receptivity of the female of either the same flower or others of the same species.

The centre of the flower is occupied by the gynoecium, which is formed of one or more carpels. Carpels may fuse to form a compound pistil or may remain separate and unfused. The carpel is divided into three regions, from the tip: the stigma, a receptive surface for pollen; the style, a specialized region for pollen tube elongation; and the ovary, a location where immature seeds called ovules occur, containing the female gametophyte or embryo sac. Meiosis occurs deep within the anther to produce many haploid male gametophytes (which become pollen) and deep within the ovule to produce the female gametophyte (or embryo sac).

The group of angiosperms known as monocotyledons typically have floral organs in multiples of three, whereas dicotyledons have floral organs in multiples of four or five. Fusion of floral organs often occurs obscuring these parts and increasing the diversity of floral form. Similar organs fuse in the example of the corolla tube (as in the morning glory), but dissimilar organs fuse in the example of apple flowers.

Organs may also be modified to perform different functions. Roses have five petals (as do most dicots), but some or all of the numerous stamens form petal-like staminodes, sterile petal-like stamens that give horticultural roses the appearance of having many more petals. Symmetry of individual flowers may be radial (forming mirror images around the centre), or bilateral (forming mirror images along only one plane).

The flower is the focus of considerable metabolic activity for the plant, producing reproductive organs with high energy content. Flowers attract

Snapdragons in bloom. Insects and other animal pollinators that visit flowers are often rewarded in return with pollen or nectar to assure this continued relationship.

insects and other animal pollinators to visit the flower. In return, pollinators are often provided with a food reward of pollen or nectar to assure this continued relationship. After the flower is fertilized, floral organs, including petals, stigma, and style, die back while seeds form inside the carpels. The tissues surrounding the ovary are dramatically modified to become the fruit.

PLANT DEVELOPMENT

Plant development is an umbrella term for a broad spectrum of processes that include: the formation of a complete embryo from a zygote; seed germination; the elaboration of a mature vegetative plant from the embryo; the formation of flowers, fruits, and seeds; and many of the plant's responses to its environment. Plant development encompasses the growth and differentiation of cells, tissues, organs, and organ systems. Plant development shares many similarities with developmental processes in animals, but the fact that plants are nonmotile, photosynthetic organisms requires certain novel developmental processes in addition to the common ones.

Embryo and Seed Development

Embryogenesis, the formation of a multicellular embryo from a single-celled zygote, is one of the most dramatic and best-characterized aspects of plant development. Four key developmental processes take place during embryogenesis. First, the zygote expresses apical-basal polarity, meaning that the apical and basal ends of the zygote cell differ structurally and biochemically. When the zygote divides, it typically divides asymmetrically, giving rise to a small apical cell with dense cytoplasm and a large basal cell with watery cytoplasm. Although these two cells have identical nuclei, their fates differ dramatically. The apical cell gives rise to the embryo itself, while the basal cell gives rise to a short-lived structure called a suspensor and the tip of the root system. The progeny of the apical cell grow.

Embryo formation begins with cell division that establishes the apical-basal (top-bottom) axis. Further divisions elaborate on this basic plan, finally forming the cotyledons (seed leaves), as well as the apical meristems of root and shoot.

spherical mass of cells, the globular-stage embryo. Second, differential growth within the globular embryo gives rise to the "heart" stage embryo, the earliest stage when the precursors of cotyledons, root, and stem can be recognized. This key embryogenic process is called organogenesis. Third, distinctive planes of cell divisions bring about histogenesis, the process by which cells within embryonic cotyledons, root, and stem acquire different shapes, forming the precursors of the plant tissue systems. Last, the apical meristems of the shoot and root systems are formed at the apical and basal ends of the embryo.

After an embryo has reached full size, developmental changes continue to occur at the cellular level. Embryonic cells, particularly those of the cotyledons, begin to synthesize and store the proteins, lipids, and starch that will provide the energy and basic building blocks for germination and seedling growth. Next, the embryo begins to desiccate, sometimes losing up to 80 percent of its previous water content, and enters a phase of dormancy. Development and metabolism are arrested in dormant embryos, and seeds containing dormant embryos can survive for many years (sometimes centuries) and withstand extreme temperatures and drought.

Plant hormones are important regulators of embryogenesis and seed dormancy. The hormones auxin, gibberellic acid, and cytokinin all stimulate growth and are present in the embryo during the stages of embryogenesis. As the embryo matures, these hormones are degraded and abscisic acid is synthesized by the embryo. Abscisic acid provides a developmental signal for the embryo to initiate the synthesis of storage compounds and to undergo desiccation. Abscisic acid is present in dormant seeds and is thought to play an important role in maintaining seed dormancy.

GERMINATION AND SEEDLING DEVELOPMENT

Embryo development and metabolism resume upon seed germination. Given the right combination of water availability, temperatures, and light, the desiccated seed begins to take up water and the embryo begins to grow and metabolize again. Some species have specific requirements for germination; for instance, many temperate zone tree species require several weeks of temperatures of 4 degrees Celsius (39.2 degrees Fahrenheit) or less in

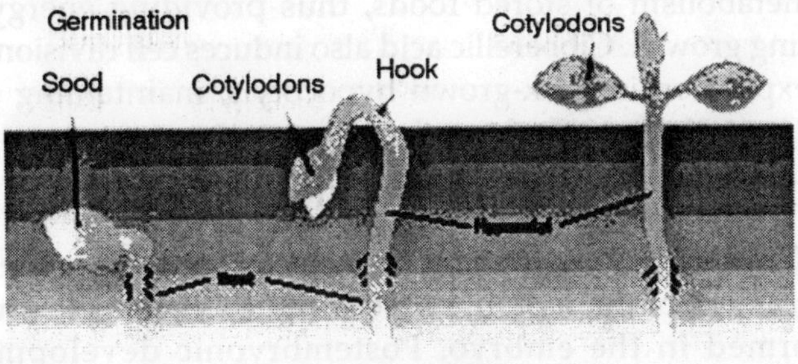

Fig. Germination of Seed

The root is the first portion of the plant to emerge during germination. Growth of the stem behind the cotyledons forms a "hook" that emerges from the soil, followed by emergence of the cotyledons, which begin to photosynthesize to feed further growth.

order to germinate. Other species require low levels of light in order to germinate. Once germination is initiated, the embryo follows a typical pattern of development. In many plants, the preformed embryonic root elongates first, forcing its way out of the seed coat and into the soil. Next, the embryonic stem, usually the part below the attachment of the cotyledons (the hypocotyl), elongates. Once the hypocotyl has carried the cotyledons into the light, they expand, providing a broad surface for photosynthesis.

Environmental factors and their translation into hormonal signals are important for seedling development. For instance, germination in the dark results in developmental events that help the seedling push its way through the soil into the light. The hypocotyl elongates quickly and maintains a "hook" near its tip that protects the cotyledons and shoot apical meristem region.

Cotyledon expansion is suppressed so that they are not damaged as they are pulled through the soil. In contrast, if the same seeds germinate in the light, the hypocotyl hardly elongates at all and does not form a hook, while the cotyledons quickly expand. The hormone gibberellic acid plays an important role in seed germination and early seedling growth. Gibberellic acid induces the synthesis of enzymes required for the metabolism of stored foods, thus providing energy for seedling growth. Gibberellic acid also induces cell division and cell expansion in dark-grown hypocotyls, maintaining their rapid growth through the soil.

APICAL MERISTEMS AND DEVELOPMENT

The early stages of germination simply involve the enlargement of the root, hypocotyl, and cotyledons that were preformed in the embryo. Postembryonic development, however, is focused on the apical meristems. The shoot apical meristem is the source of all the leaves, stems, and their component cells formed during the lifetime of the plant. The meristem itself is composed of a small population of perpetually embryonic (meristematic) cells.

These cells grow and divide, giving rise to new cells, but never mature themselves. Thus there is always a source of new cells at the tip of the shoot. The root tip has a similar population of meristematic cells that gives rise to all root tissues. Both of these meristems are characterized by an indeterminate growth pattern: one that is not finite, but, in theory at least, could continue throughout the lifetime of the plant.

Apical meristems are involved in several distinct developmental processes. The meristems are the location of cell proliferation and thus the source of all new cells in the shoot and root systems.

The regions below the meristems are the sites of active growth, as new shoot and root tissue rapidly expands. The shoot apical meristem plays a role in organogenesis, the formation of new leaves and axillary buds in a precise spatial pattern.

In contrast, the root apical meristem is not involved in organogenesis; lateral roots are initiated by pericycle cells, which are themselves derived from the meristem, usually several centimeters away from the meristem. The apical meristems also play a role in histogenesis by giving rise to cells that undergo distinct patterns of differentiation to form the specialized tissue types of the shoot and root. While the embryo initially gives rise to the precursors of dermal, ground, and vascular tissues (protoderm, ground meristem, and procambium, respectively), these tissue precursors continue to be formed by the apical meristems and represent the first stages of cell and tissue differentiation.

CELL GROWTH AND CELL DIVISION

Growth is defined as an irreversible increase in mass that is typically associated with an increase in volume. Plant cell growth is associated with meristems and must be carefully regulated in order for organogensis and histogenesis to occur in the appropriate patterns.

The plant regulates growth by regulating the extensibility of its cell walls. A cell that has nonextensible cells walls can take up some water, but eventually the physical pressure of the water inside the cell pressing out on the cell wall (the turgor pressure) prevents the entry of additional water and any further change in volume.

In contrast, a cell that has extensible cell walls can take up a substantial volume of water and thus increase in size. Turgor pressure that would otherwise prevent water entry momentarily decreases because the walls keep stretching.

Typically cell growth occurs in small increments:

• wall extensibility increases, reducing turgor pressure;

• reduced turgor pressure allows water to enter the cell, increasing cell volume;

• wall extensibility decreases, allowing the cell to build up turgor and preventing further water entry

• the cell undergoes a cycle of synthesis of cytoplasmic and wall components, adding to the cell's mass. This cycle of incremental growth is repeated many times until the cell reaches its final size.

The plant hormones auxin and gibberellin are produced in the vicinity of the apical meristems and usually act in concert to induce cell growth. Both hormones regulate wall extensibility, but carry out this function in different ways. Auxin induces the activity of cell membrane H^+ adenosine triphosphatase (ATPase) molecules. Proton (H^+) extrusion lowers the pH of the cell wall, thus activating the cell wall enzyme expansin.

Expansin cleaves the hydrogen bonds between two cell wall components: The cellulose microfibrils and the hemicellulose molecules that link adjacent cellulose microfibrils. Breakage of these bonds allows these structural wall components to reposition themselves farther apart, increasing wall extensibility. Gibberellic acid, on the other hand, stimulates the activity of another cell wall enzyme called xyloglucan endotransglycosylase (XET). Xyloglucans are a type of hemicellulose that is cleaved by the XET enzyme.

The hormones auxin and gibberellin each promote cell expansion by loosening the bonds between adjacent cell wall molecules. Each hormone acts on a different molecular target.

Hemicellulose molecules also allows the cellulose microfibrils to move farther apart, increasing wall extensibility.

Cell division and cell growth are often tightly linked. When the rate of cell division is balanced by cell growth, as in the apical meristems, average cell size does not increase. As the meristem grows away from earlier formed cells, the ratio of growth to division increases, resulting in overall cell enlargement. As the tissues mature further, cell division ceases completely, giving rise to zones of pure cell enlargement where most of the visible growth of the plant occurs. This relationship between division and growth, coupled with observations of the predictable planes of cell division during histogenesis, indicates that cell division is carefully regulated during plant development.

Molecules called cyclin-dependent kinases (CDKs) are key regulators of cell cycling (including cell division) in plants. CDKs are activated by association with a regulatory subunit called a cyclin and by phosphorylation and dephosphorylation events. The plant hormone cytokinin appears to regulate the cell cycle by interacting with the CDKs. Cytokinins enhance the synthesis of the cyclin subunits that are required for the cell to enter the deoxyribonucleic acid (DNA) synthesis phase of the cell cycle. Cytokinins also enhance the CDK dephosphorylation step that is required for the cell to progress into mitosis.

Both of these processes are inhibited by the hormone abscisic acid; thus a "developmental tug-of-war" occurs between a division-enhancing hormone and a division-suppressing hormone. The delicate balance between them determines the rate of cell division and this type of interaction is probably typical of the hormonal regulation of many aspects of plant development.

Differentiation

Differentiation is the process whereby cells, tissues, and organs become different from each other and from their precursors. The concept can be applied to organogenesis since cotyledons, foliage leaves, sepals, and petals may all develop from similar appearing precursors, the leaf primordia. As these organs mature, they become different from each other in size, shape, and the development of distinctive cell types. For instance, the epidermis tissue of petals is sharply differentiated from that of cotyledons, foliage leaves and sepals that are photosynthetic organs.

Correlated with a photosynthetic function, the epidermis of these organs is made up of flat, transparent cells that allow the penetration of light into internal tissues. Specialized guard cells that allow CO_2 to enter the leaf are also present. In contrast, the epidermal cells of petals contain brightly coloured carotenoid or anthocyanin pigments. These cells also have a papillate shape that imparts a velvetlike sheen to the petal surface. Since petals carry on minimal photosynthesis, they often lack guard cells.

The process of differentiation is best understood on a cellular level. For instance, guard cells are highly specialized epidermal cells. Early in the development of a leaf, protodermal precursor cells undergo a distinctive pattern of cell divisions. At first the cell divisions are asymmetric, producing one large and one small derivative.

The large derivative stops dividing and differentiates as an unspecialized epidermal cell, while the small derivative undergoes another asymmetric division. At an unknown stop signal, the small derivative undergoes a symmetric division, giving rise to two equal sized cells that become the guard cells. Unlike their plain neighbours, these cells develop a distinctive kidney shape, unevenly thickened cell walls, large, conspicuous chloroplasts, and finally form a pore (the stomatal aperture) between them.

UNIQUENESS OF PLANT DEVELOPMENT

Although plants share many features of development with animals such as apical-basal polarity, regulation of the balance between cell growth and cell division, formation of distinctive patterns of organs, cells and tissues, and differentiation, some aspects of development are unique to plants. Among these are:

- The formation and maintenance of the perpetually embryonic regions, the apical meristems. The meristems have an indeterminate growth pattern that result in the occurrence of growth, organogenesis, and histogenesis throughout the life of the plant.
- Plant cells have rigid cell walls that prevent cell movement. Thus organogenesis and histogenesis must occur through differential growth and regulation of the planes of cell division. Cell-cell

communication is important in plant development, but cell recognition is likely less important than it is in animals since plant cells keep the same neighbours throughout their life.

- Plant cells are totipotent; that is, able to differentiate as a different cell type if given the appropriate stimulus. Totipotency is likely a reflection of the plant's sedentary lifestyle. Plants can't escape predators and other kinds of damage, but they can readily repair wounds and reconnect vascular strands by differentiating the appropriate cell types.

FERTILIZATION

Fertilization, the process in which gametes—a male's sperm and a female's egg or ovum—fuse together, producing a single cell that develops into an adult organism. Fertilization occurs in both plants and animals that reproduce sexually—that is, when a male and a female are needed to produce an offspring. This article focuses on animal fertilization. For information on plant fertilization.

Fertilization is a precise period in the reproductive process. It begins when the sperm contacts the outer surface of the egg and it ends when the sperm's nucleus fuses with the egg's nucleus. Fertilization is not instantaneous—it may take 30 minutes in sea urchins and up to several hours in mammals. After nuclear fusion, the fertilized egg is called a zygote. When the zygote divides to a two-cell stage, it is called an embryo.

Fertilization is necessary to produce a single cell that contains a full complement of genes. When a cell undergoes meiosis, gametes are formed—a sperm cell or an egg cell. Each gamete contains only half the genetic material of the original cell. During sperm and egg fusion in fertilization, the full amount of genetic material is restored: half contributed by the male parent and half contributed by the female. In humans, there are 46 chromosomes (carriers of genetic material) in each human body cell—except in the sperm and egg, which each have 23 chromosomes. As soon as fertilization is complete, the zygote that is formed has a complete set of 46 chromosomes containing genetic information from both parents.

The fertilization process also activates cell division. Without activation from the sperm, an egg typically remains dormant and soon dies. In general, it is fertilization that sets the egg on an irreversible pathway of cell division and embryo development.

The Fertilization Process

Fertilization is complete when the sperm's nucleus fuses with the egg's nucleus. Researchers have identified several specific steps in this process. The first step is the sperm approaching the egg. In some organisms, sperm just swim randomly toward the egg (or eggs). In others, the eggs secrete a chemical substance that attracts the sperm toward the eggs. In one species of sea urchin (an aquatic animal often used in fertilization research), the sperm swim toward a small protein molecule in the egg's protective outer layer, or surface coat. In humans there is evidence that sperm are attracted to the fluid surrounding the egg.

The second step of fertilization is the attachment of several sperm to the egg's surface scoat. All animal eggs have surface coats, which are variously named the vitelline envelope (in abalone and frogs) or the zona pellucida (in mammals). This attachment step may last for just a few seconds or for several minutes.

The third step is a complex process in which the sperm penetrate the egg's surface coat. The head, or front end, of the sperm of almost all animals except fish contains an acrosome, a membrane-enclosed compartment.

The acrosome releases proteins that dissolve the surface coat of an egg of the same species.

In mammals, a molecule of the egg's surface coat triggers the sperm's acrosome to explosively release its contents onto the surface coat, where the proteins dissolve a tiny hole. A single sperm is then able to make a slitlike channel in the surface coat, through which it swims to reach the egg's cell membrane. In fish eggs that do not have acrosomes, specialized channels, called micropyles, enable a single sperm to swim down through the egg's surface coat to reach the cell membrane. When more than one sperm enters the egg, the resulting zygote typically develops abnormally.

The next step in fertilization—the fusion of sperm and egg cell membranes—is poorly understood. When the membranes fuse, a single sperm and the egg become one cell. This process takes only seconds, and it is directly observable by researchers. Specific proteins on the surface of the sperm appear to induce this fusion process, but the exact mechanism is not yet known.

After fusion of the cell membranes the sperm is motionless. The egg extends cytoplasmic fingers to surround the sperm and pull it into the egg's cytoplasm. Filaments called microtubules begin to grow from the inner surface of the egg cell's membrane inward toward the cell's centre, resembling spokes of a bicycle wheel growing from the rim inward toward the wheel's hub.

As the microtubules grow, the sperm and egg nuclei are pushed toward the egg's centre. Finally, in a process that is also poorly understood, the egg and sperm nuclear envelopes (outer membranes) fuse, permitting the chromosomes from the egg and sperm to mix within a common space. A zygote is formed, and development of an embryo begins.

Types of Fertilization

Two types of fertilization occur in animals: external and internal. In external fertilization the egg and sperm come together outside of the parents' bodies. Animals such as sea urchins, starfish, clams, mussels, frogs, corals, and many fish reproduce in this way. The gametes are released, or spawned, by the adults into the ocean or a pond. Fertilization takes place in this watery environment, where embryos start to develop.

A disadvantage to external fertilization is that the meeting of egg and sperm is somewhat left to chance. Swift water currents, water temperature changes, predators, and a variety of other interruptions can prevent fertilization from occurring. A number of adaptations help ensure that offspring will successfully be produced. The most important adaptation is the production of literally millions of sperm and eggs—if even a tiny fraction of these gametes survive to become zygotes, many offspring will still result.

Males and females also use behavioural clues, chemical signals, or other stimuli to coordinate spawning so that sperm and eggs appear in the water at the same time and in the same place. In animals that use external fertilization, there is no parental care for the developing embryos. Instead, the eggs of these animals contain a food supply in the form of a yolk that nourishes the embryos until they hatch and are able to feed on their own.

Internal fertilization takes place inside the female's body. The male typically has a penis or other structure that delivers sperm into the female's reproductive tract. All mammals, reptiles, and birds as well as some invertebrates, including snails, worms, and insects, use internal fertilization. Internal fertilization does not necessarily require that the developing embryo remains inside the female's body. In honey bees, the queen bee deposits the fertilized eggs into special compartments in the honeycomb. These compartments are supplied with food resources for the young bees to use as they develop.

Various adaptations have evolved in the reproductive process of internal-fertilizing organisms. Because the sperm and egg are always protected inside the male's and female's bodies—and are deliberately placed into close contact during mating—relatively few sperm and eggs are produced. Many animals in this group provide extensive parental care of their young. In most mammals, including humans, two specialized structures in the female's body further help to protect and nourish the developing embryo.

One is the uterus, which is the cushioned chamber where the embryo matures before birth; the other is the placenta, which is a blood-rich organ that supplies nutrients to the embryo and also removes its wastes.

Research Issues

Although reproduction is well studied in many kinds of organisms, fertilization is one of the least understood of all fundamental biological processes. Our knowledge of this fascinating topic has been vastly improved by many recent discoveries. Researchers have discovered how to clone the genes that direct the fertilization process.

Yet many important questions still remain. Scientists are actively trying to determine issues such as how sperm and egg cells recognize that they are from the same species; what molecules sperm use to attach to egg coats; and how signals on the sperm's surface are relayed inside to trigger the acrosome reaction. With continued study, answers to these questions will one day be known.

GAMETE

Gamete, sexual reproductive cell that fuses with another sexual cell in the process of fertilization. The cell resulting from the union of two gametes is called a zygote; the zygote usually undergoes a series of cell divisions until it develops into a complete organism.

Gametes, also called germ cells, vary widely in structure. The simplest sexual organisms are isogamous, that is, they produce a single kind of gamete. The identical gametes unite in pairs to produce zygotes. Although all isogametes are apparently alike in structure, they are thought to be different in physiological constitution, because gametes from the same individual do not successfully unite. The simplest isogametes, those of lower fungi such as molds, are small cells that grow on the ends of body filaments and become detached when mature. Other lower organisms, such as lower algae and protozoa, have gametes, which are formed by division of the protoplasm of single cells.

All higher plants are heterogamous, that is, they produce two kinds of gametes. The female gamete is called the egg; the male gamete is called the sperm. The organ of gamete production in plants is called a gametangium.

All animals and animal-like lower organisms that reproduce sexually, except a few protozoans, are also heterogamous. The male gametes are called spermatozoa; female gametes, ova or eggs. The gamete-producing organs of animals are called gonads. The formation of gametes in the gonads of animals is called gametogenesis. By this process the number of chromosomes in the sex cells is reduced in number from diploid to haploid, which is half the number of chromosomes in the normal body cells of the species. The diploid number of human chromosomes.

When a human sex cell divides to form two gametes, each gamete receives only half, or 23, of the normal complement of chromosomes. This type of cell division is called meiosis. The normal total of chromosomes is restored in fertilization when two gametes fuse, each contributing half of the chromosomes required by the zygote.

PLANT PROPAGATION

Plant Propagation, growing new plants from seeds or from parts of existing plants. Plant propagation occurs in nature to ensure survival and spread of species. It is also used commercially to produce seeds and plants for agriculture, horticulture, and forestry. Plant propagation includes sexual propagation, which involves the union of sperm and egg to form seeds, and asexual propagation. Asexual propagation, also known as vegetative propagation, is the growing of new plants from a leaf, stem, or root of a single parent plant. These two forms of plant propagation transmit genetic information between plants of the same species.

Genetic engineering transfers genes from one organism to another that may or may not be of the same species to introduce desirable traits into an organism. These so-called transgenic plants can then be propagated by sexual or asexual propagation.

Sexual Propagation

In nature, sexual propagation begins when water, wind, insects, birds, or small mammals carry pollen randomly between plants. In flowering plants, this transfer of pollen enables the male sex cells, or sperm, of one flower to fertilize the female sex cell, or egg, of a second flower.

The egg is located at the base of the flower in a structure called an ovule, found within the ovary. Depending on the species, an ovary contains one, several, or many ovules. The ovaries of peach and avocado flowers, have one ovule, while those of watermelon and cantaloupe have many.

As the fertilized egg (or eggs) within the ovule begins to develop into an embryonic plant, it produces a variety of hormones that stimulate the outer wall of the ovule to harden into a seed coat.

Other biochemical changes in the ovule produce a starchy substance that will be used as a food supply. In this way, the ovule ripens into a seed—a structure containing an embryonic plant and its food supply surrounded by a seed coat. The ovary, which houses the ovule or ovules, is also stimulated by hormones, which cause its tissues to enlarge into a fruit. The fruit contains the ripened ovules, or seeds.

In certain nonflowering plants such as pine, spruce, and fir trees, cones, not flowers, hold the ovules. The young cone, which averages about 2.5 cm (about 1 in) in length, is composed of numerous soft, green, leaf-like scales arranged in a spiral, with two ovules on the upper surface of each scale. Wind transports pollen to the cone scale, where it enters the ovule and fertilizes the egg. The fertilized egg develops into a seed that lies exposed on the cone scale.

Seeds, whether from flowering or nonflowering plants, typically contain a random combination of the genetic material from the sperm and egg of the parents and are therefore genetically unique. Plants grown from a seed typically have different traits from their parents—they might be weaker, with poor root development or little resistance to disease, or stronger, with good roots and successful resistance to disease. A population with diverse traits is more likely to survive changes in the environment such as emerging diseases, new predatory insects, or climate changes.

While environmental fluctuations may kill some plants, those with the necessary adaptations are able to survive, reproduce, and carry on the species. In this way, traits that help plants survive in different environments get passed down from generation to generation, resulting in the evolution of new species with traits different from those of their ancestors.

Sexual propagation also helps plants spread geographically because it produces fruits and seeds that animals such as squirrels and birds consume. The seeds pass through an animal's digestive system and are deposited at a distance from the parent plants. It is often advantageous for seeds to germinate far from their parents so that parents and offspring are not competing for light, water, and nutrients.

Seed companies take advantage of naturally pollinated plants by harvesting seed from the company fields for sale to farmers, home gardeners, and nursery workers. Seed companies, geneticists, and other researchers also use artificial pollination, or hand pollination, to control the transfer of pollen between plants.

In artificial pollination, a paintbrush or other delicate tool is used to transfer pollen between selected plants. The physical transfer of pollen in this way is called a cross, and the plants that result are also referred to as crosses. Some researchers use artificial pollination to develop new agricultural crops that combine the best traits of both parents.

A cross between wheat and rye, produced triticale, a vigourous pasture grass that resists many diseases and insects, and whose seeds are used to make a nutritious flour. Geneticists also use artificial pollination to study how particular traits are inherited within a species—to determine, if the trait for large seeds is dominant or recessive.

Asexual Propagation

Asexual propagation is the production of new plants from the leaves, stems, or roots of a single parent plant. Asexual propagation, which does not require pollination or fertilization, is a rapid method of propagation. It ensures that all of the parent's genetic material survives even if the parent dies, and it creates offspring, known as clones, with the same traits as the parent plant. Asexual propagation is advantageous when plants are well adapted to a particular environment. Several methods of asexual propagation occur in nature. They have been adapted for commercial use for rapid propagation and to obtain plants that are hard to grow from seeds.

Cuttings and Layering

For many plant species, a leaf, section of stem, or piece of root cut from a plant and lightly covered in soil, peat moss, or another growth medium develops a new, independent plant by generating the missing parts. Stimulated by hormones called auxins, a partially buried leaf or piece of stem, develops roots on the buried portion, and a piece of root forms stems and leaves above the soil. Cuttings and layering are widely used for perennial plants, plants that grow back from the same roots year after year. Commercially, cuttings are the most important source for perennials, such as new fruit trees; conifers, including pine and spruce; a variety of shrubs, roses and honeysuckle, and many florist blooms.

In layering, a new plant develops from a stem that is still attached to the parent plant. In nature, the stem simply arches over and spreads out on the ground. The parts of the stem that are in secure contact with the soil develop the roots, stems, and leaves of a new plant. The flexible stems of trailing blackberries, black raspberries, and several other species spread rapidly in this way.

Growers typically layer species that propagate naturally in this way. Layering is also used for plants such as filberts and Muscadine grapes that cannot be propagated easily by other methods. If plants do not layer naturally, a grower can induce them to layer by pinning the stems to the ground. Growers also layer plants by cutting them to the ground during the winter and covering the new spring shoots with soil, a technique called mound layering. This causes new roots to form along the buried lower portion of the new shoots, and new stems and leaves then grow up through the mound of soil. In a few weeks the new plants are cut away from the parent and planted in a garden or nursery.

This technique enables a grower to produce many plants from a single parent that has desired traits. Apple trees, currants, and gooseberries are commonly propagated in this way. In air layering, often used with house plants, stems are partially cut and peat moss or sphagnum moss is wrapped around the wound, which is then covered in plastic wrap. A single plant develops from the place where the stem was cut.

Grafting

In grafting, a freshly cut section of stem with buds, called a scion, is joined to another plant called the stock. The upper stem of the stock is severed and the scion is joined to the lower stem. The scion is securely attached to the stock, and the tissues of the two plants grow into each other, forming a single plant. The scion produces the stems, leaves, and flowers on the new plant and the stock provides the root system.

Grafting combines desirable qualities from one species, such as disease resistance or the ability to grow in waterlogged soils, with desirable qualities of another, such as the ability to produce high quality fruit. Grafting is often used to make fruit trees more vigourous and productive. Bud grafting is a form of grafting in which a single bud cut from a stem is grafted onto the stock.

It can be carried out more rapidly than other forms of grafting and is used widely in the nursery industry to propagate hundreds or thousands of plants in a relatively short amount of time. In nature, roots of oak trees of the same species commonly graft together, hastening the spread of diseases such as oak wilt, a fungal disease that kills a variety of oak trees.

Agamospermy

In agamospermy, also known as apomixis, a seed develops directly from tissues of the ovule rather than from a fertilized egg. Depending on the species, a fruit may or may not be produced. The plants that develop by agamospermy are clones of the mother plant. Agamospermy occurs in nature in species such as dandelions and blackberries, enabling them to spread rapidly since they can bypass pollination and fertilization.

Tissue Culture

Also called micropropagation, tissue culture is the production of plants under sterile laboratory conditions. A variety of tissue culture techniques are used to propagate plants. In one method, growers remove a tiny piece of leaf or stem from a plant and place it in a sterile test tube on a gel-like medium enriched with hormones and nutrients.

A yellow-brown mass of cells called callus develops from the piece of plant. Small chunks of the callus are separated, and each piece is placed in a petri dish with a hormone and nutrient mix that stimulates the development of the callus pieces into plants. The young plants are removed from the petri dish and placed in pots with soil, or into the ground, where they grow to maturity.

Tissue culture enables researchers and growers to rapidly generate numerous clones year-round in greenhouses. In nature, strawberry plants typically produce their fruits in summer. Commercially grown strawberries, however, are propagated throughout the year by tissue culture, providing consumers with a steady supply of strawberries for every season. Tissue culture is also used to produce plants free of viruses, fungi, and bacteria, and to propagate species such as Douglas fir and rhododendron, which are difficult to grow commercially from cuttings, layering, or grafting.

Propagation from Stems and Roots

Some plants produce special underground stems such as tubers, bulbs, and corms that enable them to reproduce asexually. Like all stems, these structures have buds, or nodes, from which new stems branch. Tubers are swollen, fleshy stems with several buds called eyes that produce new plants; an example of a tuber is the potato. Bulbs, such as those found in onions, lilies, hyacinths, and tulips, are short, wide, teardrop-shaped underground stems surrounded by scaly leaves. Corms, such as crocuses and gladioli, are similar, but lack the scaly leaves.

Both bulbs and corms make clumps of new bulbs or corms, called offsets, which can be divided and buried in the soil to generate new plants. Irises and ferns produce rhizomes, fleshy stems that grow horizontally beneath the soil, with new plants developing from the tip of the rhizome and from each node on the stem. Stolons, specialized stems found in strawberries and many lawn grasses, are similar to rhizomes but are usually thinner and grow on top of the soil. They also produce new plants at the tip and from the nodes.

REPRODUCTION IN PLANTS

Plant reproduction is the process by which plants generate new individuals, or offspring. Reproduction is either sexual or asexual. Sexual reproduction is the formation of offspring by the fusion of gametes. Asexual reproduction is the formation of offspring without the fusion of gametes. Sexual reproduction results in offspring genetically different from the parents. Asexual offspring are genetically identical except for mutation.

In higher plants, offspring are packaged in a protective seed, which can be long lived and can disperse the offspring some distance from the parents. In flowering plants (angiosperms), the seed itself is contained inside a fruit, which may protect the developing seeds and aid in their dispersal.

Sexual Reproduction in Angiosperms: Ovule Formation

All plants have a life cycle that consists of two distinct forms that differ in size and the number of chromosomes per cell. In flowering plants, the

large, familiar form that consists of roots, shoots, leaves, and reproductive structures (flowers and fruit) is diploid and is called the sporophyte. The sporophyte produces haploid microscopic gametophytes that are dependent on tissues produced by the flower. The reproductive cycle of a flowering plant is the regular, usually seasonal, cycling back and forth from sporophyte to gametophyte.

The flower produces two kinds of gametophytes, male and female. The female gametophyte arises from a cell within the ovule, a small structure within the ovary of the flower. The ovary is a larger structure within the flower that contains and protects usually many ovules. Flowering plants are unique in that their ovules are entirely enclosed in the ovary. The ovary itself is part of a larger structure called the carpel, which consists of the stigma, style, and ovary. Each ovule is attached to ovary tissue by a stalk called the funicle. The point of attachment of the funicle to the ovary is called the placenta.

Fig: A Hibiscus Flower, Showing anthers, five Stigmas, and pollen.

As the flower develops from a bud, a cell within an ovule called the archespore enlarges to form an embryo-sac mother cell (EMC). The EMC divides by meiosis to produce four megaspores. In this process the number of chromosomes is reduced from two sets in the EMC to one set in the megaspores, making the megaspores haploid. Three of the four megaspores degenerate and disappear, while the fourth divides mitotically three times to produce eight haploid cells. These cells together constitute the female gametophyte, called the embryo sac.

The eight embryo sac cells differentiate into two synergids, three antipodal cells, two fused endosperm nuclei, and an egg cell. The mature embryo sac is situated at the outer opening (micropyle) of the ovule, ready to receive the sperm cells delivered by the male gametophyte.

Pollen

The male gametophyte is the mature pollen grain. Pollen is produced in the anthers, which are attached at the distal end of filaments. The filament and anther together constitute the stamen, the male sex organ. Flowers usually produce many stamens just inside of the petals. As the flower matures, cells in the anther divide mitotically to produce pollen mother cells (PMC). The PMCs divide by meiosis to produce haploid microspores in groups of four called tetrads. The microspores are housed within a single layer of cells called the tapetum, which provides nutrition to the developing pollen grains.

Each microspore develops a hard, opaque outer layer called the exine, which is constructed from a lipoprotein called sporopollenin. The exine has characteristic pores, ridges, or projections that can often be used to identify a species, even in fossil pollen. The microspore divides mitotically once or twice to produce two or three haploid nuclei inside the mature pollen grain. Two of the nuclei function as sperm nuclei that can eventually fuse with the egg and endosperm nuclei of the embryo sac, producing an embryo and endosperm, respectively.

For sexual fusion to take place, however, the pollen grain must be transported to the stigma, which is a receptive platform on the top of the style, an elongated extension on top of the carpel(s). Here the moist surface or chemicals cause the pollen grain to germinate. Germination is the growth of a tube from the surface of a pollen grain. The tube is a sheath of pectin, inside of which is a solution of water, solutes, and the two or three nuclei, which lack any cell walls. Proper growth of the pollen tube requires an aqueous solution of appropriate solute concentration, as well as nutrients such as boron, which may aid in its synthesis of pectin.

At the apex of the tube are active ribosomes and endoplasmic reticulum (types of cell organelles) involved in protein synthesis. Pectinase and a glucanase (both enzymes that break down carbohydrates) probably maintain flexibility of the growing tube and aid in penetration.

The pollen tube apex also releases ribonucleic acid (RNA) and ribosomes into the tissues of the style. The tube grows to eventually reach the ovary, where it may travel along intercellular spaces until it reaches a placenta. Through chemical recognition, the pollen tube changes its direction of growth and penetrates through the placenta to the ovule. Here the tube reaches the embryo sac lying close to the micropyle, and sexual fertilization takes place.

Double Fertilization

Fertilization in flowering plants is unique among all known organisms, in that not one but two cells are fertilized, in a process called double fertilization. One sperm nucleus in the pollen tube fuses with the egg cell in the embryo sac, and the other sperm nucleus fuses with the diploid endosperm nucleus. The fertilized egg cell is a zygote that develops into the diploid embryo of the sporophyte.

The fertilized endosperm nucleus develops into the triploid endosperm, a nutritive tissue that sustains the embryo and seedling. The only other known plant group exhibiting double fertilization is the Gnetales in the genus *Ephedra,* a nonflowering seed plant. However, in this case the second fertilization product degenerates and does not develop into endosperm.

Double fertilization begins when the pollen tube grows into one of the two synergid cells in the embryo sac, possibly as a result of chemical attraction to calcium. After penetrating the synergid, the apex of the pollen tube breaks open, releasing the two sperm nuclei and other contents into the synergid. As the synergid degenerates, it envelops the egg and endosperm cells, holding the two sperm nuclei close and the other expelled contents of the pollen tube.

The egg cell then opens and engulfs the sperm cell, whose membrane breaks apart and allows the nucleus to move near the egg nucleus. The nuclear envelopes then disintegrate, and the two nuclei combine to form the single diploid nucleus of the zygote. The other sperm cell fuses with the two endosperm nuclei, forming a single triploid cell, the primary endosperm cell, which divides mitotically into the endosperm tissue.

Double fertilization and the production of endosperm may have contributed to the great ecological success of flowering plants by accelerating the growth of seedlings and improving survival at this vulnerable stage. Faster seedling development may have given flowering plants the upper hand in competition with gymnosperm seedlings in some habitats, leading to the abundance of flowering plants in most temperate and tropical regions.

Gymnosperms nevertheless are still dominant at higher elevations and latitudes, and at low elevations in the Pacific Northwest coniferous forests, such as the coastal redwoods. The reasons for these patterns are still controversial.

THE SEED V

The seed is the mature, fertilized ovule. After fertilization, the haploid cells of the embryo sac disintegrate. The maternally derived diploid cells of the ovule develop into the hard, water-resistant outer covering of the seed, called the testa, or seed coat. The diploid zygote develops into the embryo, and the triploid endosperm cells multiply and provide nutrition. The testa usually shows a scar called the hilum where the ovule was originally attached to the funicle. In some seeds a ridge along the testa called the raphe shows where the funicle originally was pressed against the ovule. The micropyle of the ovule usually survives as a small pore in the seed coat that allows passage of water during germination of the seed.

In some species, the funicle develops into a larger structure on the seed called an aril, which is often brightly coloured, juicy, and contains sugars that are consumed by animals that may also disperse the seed (as in nutmeg, arrowroot, oxalis, and castor bean). This is distinct from the fruit, which forms from the ovary itself.

The embryo consists of the cotyledon(s), epicotyl, and hypocotyl. The cotyledons resemble small leaves, and are usually the first photosynthetic organs of the plant. The portion of the embryo above the cotyledons is the epicotyl, and the portion below is the hypocotyl. The epicotyl is an apical meristem that produces the shoot of the growing plant and the first true leaves after germination. The hypocotyl develops into the root.

Often the tip of the hypocotyl, the radicle, is the first indication of germination as it breaks out of the seed.

Flowering plants are classified as monocotyledons or dicotyledons (most are now called eudicots) based on the number of cotyledons produced in the embryo. Common monocotyledons include grasses, sedges, lilies, irises, and orchids; common dicotyledons include sunflowers, roses, legumes, snapdragons, and all nonconiferous trees.

The endosperm may be consumed by the embryo, as in many legumes, which use the cotyledons as a food source during germination. In other species the endosperm persists until germination, when it is used as a food reserve.

In grains such as corn and wheat, the outer layer of the endosperm consists of thick-walled cells called aleurone, which are high in protein.

THE FRUIT

The fruit of a flowering plant is the mature ovary. As seeds mature, the surrounding ovary wall forms a protective structure that may aid in dispersal. The surrounding ovary tissue is called the pericarp and consists of three layers. From the outside to inside, these layers are the exocarp, mesocarp, and endocarp. The exocarp is usually tough and skinlike. The mesocarp is often thick, succulent, and sweet. The endocarp, which surrounds the seeds, may be hard and stony, as in most species with fleshy fruit, such as apricots.

A fruit is termed simple if it is produced by a single ripened ovary in a single flower (apples, oranges, apricots). An aggregated fruit is a cluster of mature ovaries produced by a single flower (blackberries, raspberries, strawberries). A multiple fruit is a cluster of many ripened ovaries on separate flowers growing together in the same inflorescence (pineapple, mulberry, fig). A simple fruit may be fleshy or dry. A fleshy simple fruit is classified as a berry (grape, tomato, papaya), pepo (cucumber, watermelon, pumpkin), hesperidium (orange), drupe (apricot), or pome (apple).

Dry simple fruits have a dry pericarp at maturity. They may or may not dehisce, or split, along a seam to release the seeds. A dehiscent dry fruit is classified as legume or pod (pea, bean), silique or silicle (mustard), capsule (poppy, lily), or follicle (milkweed, larkspur, columbine). An indehiscent dry fruit that does not split to release seeds is classified as an achene (sunflower, buttercup, sycamore), grain or caryopsis (grasses such as corn, wheat, rice, barley), schizocarp (carrot, celery, fennel), winged samara (maple, ash, elm), nut (acorn, chestnut, hazelnut), or utricle (duckweed family). Some fruiting bodies contain non-ovary tissue and are sometimes called pseudocarps. The sweet flesh of apples and pears, is composed not of the pericarp but the receptacle, or upper portion, of the flowering shoot to which petals and other floral organs are attached.

Fruiting bodies of all kinds function to protect and disperse the seeds they contain. Protection can be physical (hard coverings) or chemical (repellents of seed predators). Sweet, fleshy fruits are attractive food for birds and mammals that consume seeds along with the fruit and pass the seeds intact in their fecal matter, which can act as a fertilizer. Dry fruits are usually adapted for wind dispersal of seeds, with the assistance of winglike structures or a fluffy pappus that provides buoyancy. The diversity of fruiting bodies reflects in part the diversity of dispersal agents in the environment, which select for different fruit size, shape, and chemistry.

POLLINATION AND POLLINATORS

Pollination is the movement of pollen from the stamens to the stigma, where germination and growth of the pollen tube occur. Most (approximately 96 percent) of all flowering plant species are hermaphroditic (possess both sexual functions within a plant, usually within every flower), and thus an individual can be pollinated by its own pollen or by pollen from another individual. Seed produced through self-pollination ("selfed" seed) is often inferior in growth, survival, and fecundity to seed produced through outcross pollination ("outcrossed" seed). As a result, in most species there is strong natural selection to maximize the proportion of outcrossed seed (the "outcrossing rate").

Flowering plants are unusual among seed plants in their superlative exploitation of animals (primarily insects) as agents of outcross pollination. The outcross pollination efficiency of insects, birds, and mammals (primarily bats) may have contributed to both the abundance and diversity of flowering plants. Abundance may have increased because of less wastage of energy and resources on unsuccessful pollen and ovules. Diversity may have increased for two reasons.

First, insects undoubtedly have selected for a wide variety of floral forms that provide different rewards (pollen and nectar) and are attractive in appearance (colour juxtaposition, size, shape) and scent (sweet, skunky) in different ways to different pollinators. Second, faithfulness of pollinators to particular familiar flowers may have reduced hybridization and speeded evolutionary divergence and the production of new species.

Although flowering plants first appeared after most of the major groups of insects had already evolved, flowering plants probably caused the evolution of many new species within these groups. Some new insect groups, such as bees and butterflies, originated after flowering plants, their members developing mouthpart structures and behaviour specialized for pollination.

In extreme cases, a plant is completely dependent on one insect species for pollination, and the insect is completely dependent on one plant species for food. Such tight interdependency occurs rarely but is well documented in yuccas/yucca moths, senita cacti/senita moths, and fig trees/some fig wasps. In all three insects, females lay eggs in the flowers, and their young hatch later to feed on the mature fruit and its contents. Females ensure that the fruit develops by gathering pollen from another plant and transporting it to the stigma of the flower holding their eggs.

Plants benefit greatly in outcrossed seed produced, at the small cost of some consumed fruit and seeds, and the insects benefit greatly from the food supply for developing larvae at the small cost of transporting pollen the short distances between plants.

Pollinating agents, whether biotic or abiotic, have exerted strong selection on all aspects of the flower, resulting in the evolution of tremendous floral diversity. This diversity has been distilled into a small number of characteristic pollination syndromes.

Pollination by beetles selects usually for white colour, a strong fruity scent, and a shallow, bowl-shaped flower. Bees select for yellow or blue/purple colourings, a landing platform with colour patterns that guide the bee to nectar (often reflecting in the ultraviolet range of the spectrum), bilateral symmetry, and a sweet scent. Butterflies select for many colours other than yellow, a corolla (petal) tube with nectar at the base, and the absence of any scent.

Moths in contrast select for nocturnally opening flowers with a strong scent and drab or white colour, and also a tube with nectar at the base. Bats select also for nocturnally opening flowers, but with a strong musky scent and copious nectar, positioned well outside the foliage for easy access, and drab or white colour.

Hummingbirds select for red or orange flowers with no scent, copious nectar production, and a corolla tube with nectar at the base. Other pollinating birds that do not hover while feeding select for strong perches and flowers capable of containing copious nectar (tubes, funnels, cup shapes).

Wind as a pollinating agent selects for lack of colour, scent, and nectar; small corolla; a large stigmatic surface area (usually feathery); abundantly produced, buoyant pollen; and usually erect styles and limp, hanging stamens. In addition there is great floral diversity within any of these syndromes, arising from the diverse evolutionary histories of the member plant species.

Selfing and Outcrossing

Most flowering plant species reproduce primarily by outcrossing, including the great majority of trees, shrubs, and perennial herbs. Adaptations that prevent self-fertilization include self-incompatibility (genetic recognition and blocking of self-pollen) and dioecy (separate male and female individuals). Adaptations that reduce the chances of self-pollination in hermaphrodites include separation of the anthers and stigma in space (herkogamy) or time (dichogamy).

In many species, both self-incompatibility and spatiotemporal separation of the sex organs occur.

The ability to produce seeds by selfing, however, is advantageous in situations where outcrossing pollination is difficult or impossible. These include harsh environments where pollinators are rare or unpredictable, and regularly disturbed ground where survivors often end up isolated from each other. Selfing is also cheaper than outcrossing, because selfers can become pollinated without assistance from animals and therefore need not produce large, attractive flowers with abundant nectar and pollen rewards.

Most primarily selfing species are small annuals in variable or disturbed habitats, with small, drab flowers. Most desert annuals and roadside weeds are selfers. The evolutionary transition from outcrossing to near-complete selfing has occurred many times in flowering plants.

Outcrossing and selfing species differ in their evolutionary potential. Outcrossers are generally more genetically diverse and produce lineages that persist over long periods of evolutionary time, during which many new species are formed. Selfers, however, are less genetically diverse and tend to accumulate harmful mutations. They typically go extinct before they have an opportunity to evolve new species.

ASEXUAL REPRODUCTION

The ability to produce new individuals asexually is common in plants. Under appropriate experimental conditions, nearly every cell of a flowering plant is capable of regenerating the entire plant. In nature, new plants may be regenerated from leaves, stems, or roots that receive an appropriate stimulus and become separated from the parent plant. In most cases, these new plants arise from undifferentiated parenchyma cells, which develop into buds that produce roots and shoots before or after separating from the parent.

New plants can be produced from aboveground or belowground horizontal runners (stolons of strawberries, rhizomes of many grasses), tubers (potato, Jerusalem artichoke, dahlia), bulbs (onion, garlic), corms (crocus, gladiola), bulbils on the shoot (lily, many grasses), parenchyma cells in the leaves (Kalanchoe, African violet, jade plant) and inflorescence (arrowhead). Vegetative propagation is an economically important means of replicating valuable agricultural plants, through cuttings, layering, and grafting.

Vegetative reproduction is especially common in aquatic vascular plants (example, surfgrass and eelgrass), from which fragments can break off, disperse in the current, and develop into new whole plants.

A minority of flowering plants can produce seeds without the fusion of egg and sperm (known as parthenocarpy or agamospermy). This occurs when meiosis in the ovule is interrupted, and a diploid egg cell is produced, which functions as a zygote without fertilization. Familiar examples include citrus, dandelion, hawkweed, buttercup, blackberry/raspberry, and sorbus. Agamospermous species are more common at high elevations and at high latitudes, and nearly all have experienced a doubling of their chromosome number (tetraploidy) in their recent evolutionary history. These species experience evolutionary advantages and disadvantages similar to those of selfers.

Evolutionary Significance of Plant Reproduction Strategies

The attractive, colourful, and unique features of the most abundant and diverse group of land plants—the flowering plants—are believed to have evolved primarily to maximize the efficiency and speed of outcross reproduction. Each major burst of angiosperm evolution was a coevolutionary episode with associated animals, primarily insects, which were exploited to disperse pollen and seeds in ever more efficient and diverse ways.

The first major burst of flowering plant evolution was the appearance of the closed carpel together with showy flowers that were radially symmetrical. The closed carpel prevented self-fertilization through recognition and blocking of self pollen within the specialized conducting tissue of the style. Insects attracted to the showy flowers carried pollen between plants less wastefully than wind, and the radial symmetry accommodated insects of many sizes and shapes.

The second major burst was the appearance of bilaterally symmetrical flowers, which happened independently in many groups of plants at the same time that bees evolved. Bilateral symmetry forced bees to enter and exit flowers more precisely, promoting even more efficient outcross pollen transfer.

The third major burst of flowering plant evolution was the appearance of nutritious, fleshy fruits and seeds, coincident with a diversification of birds and rodents. The exploitation of vertebrates for fruit and seed dispersal resulted in less haphazard transport of offspring to neighbouring populations of the same species (also visited as a food source), thereby reducing the chances that progeny inbreed with their siblings and parents and providing more assurance than wind currents that they find good habitat and unrelated mating partners of the same species.

Chapter 4

Thallus

THALLUS ORGANIZATION

Being algae, the plant body in corallines is referred to as a thallus (pl. thalli). Most coralline algae have an internal anatomy of the thallus that is based on the aggregation of branching filaments, each of which is formed through the action of its own single meristematic cell.

This type of thallus anatomy is said to be pseudoparenchymatous. Only certain parasitic species, and some epiphytes, produce filaments that may not be aggregated into pseudoparenchyma. Since their anatomy is based on the elaboration of filaments, true tissues do not occur in coralline algae, and it is therefore incorrect to use the term 'tissue' do describe regions within the thallus.

Most nongeniculate corallines show some degree of dorsiventral arrangement, that is, an arrangement in which a distinct upper and lower surface can be distinguished. There are two different types of organization in dorsiventral thalli:

- Dimerous organization;
- Monomerous organization.

In dimerous thalli, there are two distinct groups of filaments oriented more-or-less at right angles to one another, with a ventral layer of filaments (called basal filaments) that is uually one cell thick, and an erect layer in which filaments (called erect filaments) vary from one to many cells in length and comprise most of the thallus thickness.

Dimerous thalli arise when a spore germinates to produce a unistratose layer of radiating, repeatedly branched filaments, that grow via meristematic cells that are located at the apices of the branches.

The growth of this layer of basal filaments contributes to the areal expansion of the thallus. Erect filaments are formed in most dimerous species when cells of the basal filaments divide periclinally to form erect filaments.

The first cell that is cut off remains meristematic and forms new cells of the erect filament ventrally, and epithallial cells dorsally. Since it is usually overlain by one or more epithallial cells, the meristematic cell is usually referred to as a subepithallial initial. Dimerous thalli occur in *Lithophyllum, Titanoderma, Exilicrusta, Melobesia,* and some species of *Hydrolithon* and *Pneophyllum.*

In monomerous thalli, there is but a single pseudoparenchymatous system of repeatedly branched filaments in which some derivatives contribute to a medulla which runs more-or-less parallel to the thallus surface (or substratum in dorsiventral, attached species) and some derivatives curve outwards and collectively form a cortex. In most dorsiventral thalli, downwards-curving filaments are suppressed, and only occur vestigially.

The cortical filaments terminate at the upper surface in one or more epithallial cells, that are subtended by a subepithallial initial. Continued division of the subepithallial initial increases the thallus girth, and may give rise to surface protuberances which may themselves branch.

Monomerous thalli occur in *Clathromorphum, Leptophytum, Lithothamnion, Phymatolithon, Mesophyllum,* and many species of *Hydrolithon, Spongites,* and *Pneophyllum.*

In some species, the primary thallus is dimerous, but monomerous regions may develop secondarily from the cells of erect filaments (e.g. *Lithophyllum incrustans*).

In both monomerous and dimerous forms, the cells of adjacent filaments below the merismetic cells often fuse together, or form pit connections, allowing for cellular interchange throughout the plant in spite of the calcified cell walls. Cell fusions and secondary pit connections are used for taxonomic purposes at the sub-family as well as species level.

THALLOPHYTE

Thallophyte, common name for a group of organisms that were formerly placed in the division Thallophyta of the plant kingdom. They have since been reclassified, variously, as algae and fungi. Thallophytes are characterized by the simple nature of their structure, an undifferentiated thallus that has no root, stem, or leaves.

ALGAE

Algae, diverse group of simple, plantlike organisms. Like plants, most algae use the energy of sunlight to make their own food, a process called photosynthesis. However, algae lack the roots, leaves, and other structures typical of true plants. Algae are the most important photosynthesizing organisms on Earth.

They capture more of the sun's energy and produce more oxygen (a byproduct of photosynthesis) than all plants combined. Algae form the foundation of most aquatic food webs, which support an abundance of animals.

Algae vary greatly in size and grow in many diverse habitats. Microscopic algae, called phytoplankton, float or swim in lakes and oceans. Phytoplankton are so small that 1000 individuals could fit on the head of a pin. The largest forms of algae are seaweeds that stretch 100 m (300 ft) from the ocean bottom to the water's surface.

Although most algae grow in fresh water or seawater, they also grow on soil, trees, and animals, and even under or inside porous rocks, such as sandstone and limestone. Algae tolerate a wide range of temperatures and can be found growing in hot springs, on snow banks, or deep within polar ice.

Algae also form mutually beneficial partnerships with other organisms. Algae live with fungi to form lichens—plantlike or branching growths that form on boulders, cliffs, and tree trunks.

Algae called *zooxanthellae* live inside the cells of reef-building coral. In both cases, the algae provide oxygen and complex nutrients to their partner, and in return they receive protection and simple nutrients. This arrangement enables both partners to survive in conditions that they could not endure alone.

The earliest life-forms on this planet are thought to be early ancestors of cyanobacteria, a type of algae formerly called blue-green algae. Fossilized cyanobacteria have been found in rocks more than 3 billion years old. These early algae formed when there was no oxygen in the atmosphere, and scientists theorize that as the algae photosynthesized, they released oxygen as a byproduct, which eventually accumulated in the atmosphere. Algae were probably the first organisms capable of photosynthesis and, until the appearance of plants on earth, the only photosynthesizers for billions of years.

Physical Characteristics

With the exception of the cyanobacteria, algae are eukaryotes—that is, the insides of their cells are organized into separate membrane-wrapped organelles, including a nucleus and mitochondria. An important organelle found in eukaryotic algae is the chloroplast, which contains the light-absorbing pigments responsible for capturing the energy in sunlight during photosynthesis. In most algae the primary pigment is chlorophyll, the same green pigment used in plants. Many algae also contain secondary pigments, including the carotenoids, which are brown or yellow, and the phycobilins, which are red or blue. Secondary pigments give algae their colourful hues. The cyanobacteria are prokaryotes—that is, relatively simple unicellular organisms lacking a nucleus and other membrane-bound organelles. As their modern name implies, the cyanobacteria have many characteristics that resemble bacteria.

Like plants, most algae have rigid cell walls composed largely of cellulose. An exception is the diatom, whose cell wall is composed primarily of silica, which provides rigidity and also produces elaborately sculpted patterns of grooves that serve as identifying features. Many eukaryotic algae have whiplike appendages called flagella attached to their cell walls. By beating flagella in a rotary movement, these algae are able to move through water with considerable speed.

A few algae that are devoid of rigid cell walls are able to protrude one part of the body ahead of the other to crawl on solid surfaces in an amoeba-like fashion.

Algae come in a variety of shapes and forms. The simplest form is the single, self-sufficient cell, such as *Euglena*, dependent only on sunlight and carbon dioxide and minerals from the water. Numerous one-celled algae may clump together to form a colony. Although these cells are attached to one another, each cell within a colony continues to function independently. Still other algae are multicellular organisms.

In the simplest multicellular algae, the cells are joined end to end, forming filaments, both branched and unbranched. More complex structures may be shaped like a small disc, tube, club, or even a tree. The most complex algae have highly specialized cells. Some seaweeds, for instance, have a variety of specialized tissues, including a rootlike *holdfast*, a *stipe*, which resembles a plant stalk, and a leaflike *blade*.

While most algae create their own food through photosynthesis, some are unable to photosynthesize. These algae ingest food from external sources by absorbing simple nutrients through the cell membrane. To absorb more complex nutrients, algae that lack rigid walls are able to engulf food particles and digest them.

Some of the algae known as dinoflagellates extend a feeding tube, called a peduncle, to suck in food. Other dinoflagellates use special harpoonlike structures to snare their food. Some algae are parasites, living in or on another organism from which they get their food. Some parasitic red algae live off other red algae, and parasitic dinoflagellates live in the intestines of some marine animals, such as copepods and annelids.

Reproduction

Algae reproduce in astoundingly diverse ways. Some reproduce asexually, others use sexual reproduction, and many use both. In asexual reproduction an individual reproduces without combining its genetic material with that from another individual. The simplest form of asexual reproduction is binary fission, in which a unicellular organism simply divides into two new individuals. Some multicellular algae, including *Sargassum*, reproduce asexually through fragmentation, in which fragments of the parent develop into new individuals. In a similar process called budding, special buds detach from multicellular algae and develop into new individuals, commonly found in *Sphacelaria*. Many algae produce special cells called spores that are capable of growing into new individuals. If these spores move about using flagella, they are known as zoospores.

In sexual reproduction, genetic material from two individuals is combined. The simplest form of sexual reproduction in algae is conjugation, in which two similar organisms fuse, exchange genetic material, and then break apart. In *Spirogyra*, which produces both asexually and sexually, two long, unbranched filaments join via conjugation tubes, through which genetic material is exchanged between cells. Most multicellular algae undergo a more complex form of sexual reproduction involving the union of special reproductive cells, called gametes, to form a single cell, known as a zygote.

Many algae incorporate both sexual and asexual modes of reproduction. This is well demonstrated in the life cycle of the alga *Chlamydomonas*. The mature alga is a single haploid cell—that is, it contains only one set of chromosomes. During asexual reproduction the cell undergoes mitosis, a type of cell division that produces genetically identical offspring. Four daughter cells are created that emerge from the enclosing parent cell as spores. The spores develop into mature haploid cells that are genetically identical to the parent cell.

Certain environmental conditions, such as lack of nutrients or moisture, may trigger the haploid daughter cells to undergo sexual reproduction. Instead of forming into spores, the haploid daughter cells form gametes that have two different mating strains. These two strains are structurally similar and are called plus and minus strains. Opposite mating strains fuse in a process known as isogamy to form a diploid zygote, which contains two sets of chromosomes. After a period of dormancy, the zygote undergoes meiosis, a type of cell division that reduces the genetic content of a cell by half. This cell division produces four genetically unique haploid cells that eventually grow into mature cells.

Some multicellular green algae, such as *Ulva*, follow a distinct pattern of reproduction called alternation of generations, in which it takes two generations—one that reproduces sexually and one that reproduces asexually—to complete the life cycle. The two mature forms of the algae, alternating between diploid and haploid individuals, are identical in appearance, or isomorphic. The haploid form, called a gametophyte, undergoes mitosis to produce haploid gametes. These gametes unite to form a diploid zygote, which develops into the diploid form called a sporophyte. The sporophyte undergoes meiosis to form haploid spores that, in turn, form gametophytes.

Not all algae that undergo alternation of generations have haploid and diploid forms that look alike. In the life cycle of the seaweed *Laminaria*, the gametophyte and the sporophyte are distinct in appearance, or heteromorphic. The *Laminaria* sporophyte appears as long, bladelike structures that grow on rocks just below the water in intertidal zones. The gametophyte is short, with branched filaments.

Types of Algae

The most common classification system distributes algae in more than one kingdom. Most algae are classified in the Kingdom Protista, along with other eukaryotic organisms that lack true specialized tissues. The cyanobacteria, however, are classified with the bacteria in the Kingdom Prokaryotae, which consists of prokaryotic organisms. This classification system continues to be intensely debated as new research increases our understanding of the way that these organisms are related.

Green Algae

Green algae form the phylum Chlorophyta and are named for their green chloroplasts, which are similar in composition to the chloroplasts found in land plants. Green algae range in shape from unicellular plankton that grow in lakes and oceans to colonial filaments of pond scum to leaflike seaweeds that grow along rocky and sandy intertidal areas. Some green algae also live on tree trunks and soil. Several green algal species are symbiotic, forming lichens with fungi or living with corals. Green algae may also be found inside freshwater sponges, giving the sponges a bright green colour, and in permanent snow banks, where a secondary pigment masks the chlorophyll and turns the snow a reddish colour.

More than 500 genera and 8000 species of green algae have been identified. Some familiar green algae include the genus *Spirogyra,* known for its spiral-shaped chloroplasts, and the desmids, recognized by their characteristic shape—two symmetrical halves, joined by a small bridge.

The green algae known as Stoneworts often grow several feet in length. Their name comes from calcium crusts that make them feel like stone.

Most green algae reproduce both sexually and asexually. Alternation of generatic ıs, where the algae alternates between gametophyte and sporophyte generations, is common among the multicellular green algae.

Red Algae

Red algae form the phylum Rhodophyta with approximately 500 genera and 6000 species. Found in warm coastal waters and in water as deep as 260 m (850 ft), red algae species adapt to varied water depths by having different proportions of pigments.

Their red colour is due to a red pigment, phycoerythrin, which is well suited to absorb the blue light that penetrates deeper into water than the other colours of light.

Red algae found in deep water may be almost black due to a high concentration of phycoerythrin. At moderate depths red algae appear red, while in shallow water they may appear green because a smaller proportion of phycoerythrin is unable to mask the green of chlorophyll. Most red algae are multicellular and come in a variety of shapes, including filaments, which are shaped like a blade of grass, and seaweed shapes. Unlike most other eukaryotic algae, red algae have no flagella. Red algae use diverse strategies to reproduce, including fragmentation and spore production.

One unusual strategy, found in many species including those in the genus *Polysiphonia*, involves the alternation among three generations. A diploid sporophyte produces diploid spores that germinate into another diploid sporophyte that looks completely different from the first one. Meiosis occurs in the second sporophyte, producing haploid spores that germinate into gametophytes. Surprisingly, in some species, the gametophytes look nearly identical to the second sporophyte.

Almost all red algae live in marine habitats, although some species are found in fresh water or damp soil. Many types of seaweed are red algae, typically found growing along the coast and attaching firmly to the seafloor using a rootlike holdfast. In some species, called coralline algae, the cell walls become hardened with calcium carbonate. Coralline algae are important members of coral reefs, producing new material and cementing together other organisms.

Golden-Brown Algae, Brown Algae, and Diatoms

Golden-brown algae, brown algae, and diatoms form the large and complex phylum Heterokontophyta, with organisms ranging in size from a fraction of a millimeter to more than 100 m (300 ft) long. Heterokontophyta have carotenoid secondary pigments that tend to mask the green of the primary chlorophyll pigment, giving them a golden or golden-brown appearance. Flagellated cells in this phylum have two types of flagella: One is smooth, while the other has two rows of stiff hairs running down opposite sides of the flagellum. Algae in this phylum typically have an eyespot that can detect light.

The golden-brown algae, also known as the yellow-brown algae, include about 200 genera and 1000 species that receive their characteristic colouring from the carotenoid pigment fucoxanthin. These algae are mostly unicellular or colonial, swimming or floating in lakes and oceans as phytoplankton. In shallow ponds that dry up in summer or freeze completely in winter, golden-brown algae survive by forming protective cysts that can withstand the harsh conditions. When favorable conditions return, the algae emerge from the cysts. Like so many other algae, the unicellular algae tend to reproduce through fission, while the multicellular and colonial forms reproduce either through fragmentation or through spore production.

Diatoms are best known for their glasslike cell wall made of silica. The cell wall has ornate ridge patterns. A diatom consists of two overlapping halves that fit together like a shoebox or a petri dish, with the lid slightly larger and fitting over the base. During asexual cell division, the two glass walls separate and serve as the lids for two new glass bases. The new diatom that grew from the lid is the same size as the parent diatom, while the diatom that grew from the smaller base is slightly smaller than its parent. Sexual reproduction occurs when the succeeding generations shrink to a critical size. These smallest diatoms form gametes that shed their glass walls. Upon fertilization, the zygotes absorb water to swell and then secrete new, full-sized silica coverings.

A very large class with more than 250 genera and 8000 species, diatoms are found floating in freshwater and seawater, growing attached to the seafloor, or growing on soil. The cells are either unicellular or form colonial chains of round cells. When an organism dies, its silica cell wall remains intact. Over time these shells have accumulated to form layers of soft rock in some geologic formations. This diatomaceous earth is mined and quarried for use in filters and bleaching agents, as an abrasive powder for cleaning and polishing metals, and for insect pest control (the broken cell walls of silica tear the insect gut).

Brown algae include over 260 genera and 1500 species. Multicellular algae, they may range from tiny filaments to the largest and most complex algae, such as the kelps, with leaflike blades and stems that can be up to 100 m (300 ft) long. Most brown algae grow in marine waters near the coast, attached to rocks either along the shoreline or underneath the ocean surface. One type, *Sargassum*, forms huge floating masses in the middle of the Sargasso Sea. The brown or olive colour is due to the pigment fucoxanthin. The life cycles of brown algae vary considerably, but most demonstrate alternation of generations.

Dinoflagellates

Dinoflagellates of the phylum Dinoflagellata are mostly unicellular organisms that may be covered with stiff cellulose plates that resemble armored helmets.

Many species have unusual ornamentation, such as horns, spines, or wings. A narrow groove encircles the armor, and a second groove runs perpendicular to the first groove. Flagella beat within these grooves, causing the dinoflagellates to spin like tops as they move through the water. Most of the 130 genera and 2000 species in this phylum are planktonic and live in saltwater, although there are many freshwater planktonic representatives as well.

Many dinoflagellate species lack chloroplasts and are dependent on other species for their food. Some are parasites, but most are carnivores, using special harpoonlike structures called trichocysts to capture other organisms to eat.

In contrast, several of the photosynthetic species live inside the tissues of invertebrate animals, such as corals and giant clams. These dinoflagellates share the food they photosynthesize with their host, and in return, receive protection and some nutrients.

Under favorable environmental conditions, some dinoflagellate species experience population explosions, known as blooms. If the species involved in the bloom have red pigments, their concentration can be high enough to turn the seawater red, forming red tides. Dinoflagellate blooms can be quite destructive.

During the night when photosynthesis halts, such a high concentration of individuals can deplete the oxygen in the water, suffocating fish.

Some dinoflagellates release toxins, some of which kill fish, while other toxins are passed up the food chain until they reach humans, where they can cause paralytic shellfish poisoning and ciguatera fish poisoning. Recently, the dinoflagellate *Pfiesteria piscicida* has caused fish, shellfish, and human disease in estuaries of the southeastern United States.

Cyanobacteria

Unlike other algae, the cyanobacteria are prokaryotes—single-celled organisms with characteristics that cause biologists to debate whether they are really algae or bacteria. Cyanobacteria are found nearly everywhere, occurring in typical aquatic and terrestrial habitats as well as in such extreme sites as hot springs with temperatures as high as 71° C (160° F) and crevices of desert rocks. Cyanobacteria make up the phylum Cyanophyta, and this phylum contains about 150 genera and 2000 species worldwide.

Like other bacteria, cyanobacteria do not have organelles such as nuclei, mitochondria, or chloroplasts. Cyanobacteria are distinguished from bacteria by the presence of internal membranes, called thylakoids, that contain chlorophyll and other structures involved in photosynthesis. While higher plants have two kinds of chlorophyll, called a and b, cyanobacteria contain only chlorophyll a. Cyanobacteria colour varies from blue-green to red or purple and is determined by the proportions of two secondary pigments, c-phycocyanin (blue) and c-phycoerythrin (red), which tend to mask the green chlorophyll present in the thylakoids.

Cyanobacteria reproduce asexually by binary fission, spore production, or fragmentation, forming singular cells, colonies, filaments, or gelatinous masses. Although most lack flagella and are nonmotile, filamentous forms such as *Oscillatoria* rotate in a screwlike manner, and the gelatinous forms glide along their slimy mucus.

Cyanobacteria may be both beneficial and harmful to humans: Some act as natural fertilizers in some habitats, especially rice paddies, whereas others produce toxins. Mild cyanobacteria toxins in lakes and oceans cause a rash known as swimmer's itch, while powerful neuromuscular toxins released by other cyanobacteria can kill fish living in the water or the animals that drink the water. In certain conditions, cyanobacteria may form dense blooms, which may produce toxins that make seafood poisonous to humans. Even if the cyanobacteria do not produce toxins, blooms can cause water to have an unpleasant taste and odor.

Algae Uses

Human ingenuity has found many uses for algae. Algae provide food for people and livestock, serve as thickening agents in ice cream and shampoo, and are used as drugs to ward off diseases. More than 150 species of algae are commercially important food sources, and over $2 billion of seaweed is consumed each year by humans, mostly in Japan, China, and Korea. The red alga *Porphyra,* called nori, is the most popular food product. After harvesting, nori is dried, pressed into sheets, and used in soups, sauces, sushi, and condiments. Algae are considered nutritious because of their high protein content and high concentrations of minerals, trace elements, and vitamins. The high iodine content of many edible algae may contribute to the low rates of goiter observed in countries where people frequently eat algae.

In coastal areas of North America and Europe, seaweeds are fed to farm animals as a food supplement. Cyanobacteria species that are high in protein, such as *Spirulina,* are grown commercially in ponds and used mostly as a health food and cattle dietary supplement. Seaweeds also are applied to soils as a fertilizer and soil conditioner, as their high concentrations of potassium and trace elements improve crop production. Some species of cyanobacteria can turn atmospheric nitrogen into ammonia, a form that can then be used by plants as a nutrient. Farmers in tropical countries grow cyanobacteria in their flooded rice paddies to provide more nitrogen to the rice, increasing productivity as much as tenfold.

Seaweeds are a critical source of three chemical extracts used extensively in the food, pharmaceutical, textile, and cosmetic industries. Brown algae yield alginic acid, which is used to stabilize emulsions and suspensions; it is found in products such as syrup, ice cream, and paint. Different species of red algae provide agar and carrageenan, which are used for the preparation of various gels used in scientific research.

Bacteria, fungi, and cell cultures are commonly grown on agar gels. Agar is also used in the food industry to stabilize pie fillings and preserve canned meat and fish. Carrageenan is also used as a thickening and stabilizing agent in products such as puddings, syrups, and shampoos.

Algae have been used for centuries, especially in Asian countries, for their purported powers to cure or prevent illnesses as varied as cough, gout, gallstones, goiter, hypertension, and diarrhea. Recently, algae have been surveyed for anticancer compounds, with several cyanobacteria appearing to contain promising candidates. Diatoms also have been used in forensic medicine, as their presence in the lungs can indicate a person died due to drowning.

Algae can also serve as indicators of environmental problems in aquatic ecosystems. Because algae grow quickly and are sensitive to changing environmental conditions, they are often among the first organisms to respond to changes. Depletion of the diatom community in the Florida Everglades provided strong evidence of phosphorus-related changes in this unique ecosystem. Algal blooms may deplete oxygen concentrations in water and smother fish and plant life, as well as prevent light penetration for algae at lower depths, preventing photosynthesis.

LICHEN

Lichen, living partnership of a fungus and an alga. The fungus component is called the mycobiont and is composed of intertwined, threadlike fibres called hyphae that are tightly packed into a tissuelike sheet. The fungus uses these hyphae to absorb food from its surroundings. The algal component, called the photobiont, makes its own food through photosynthesis and grows as a mass of green cells dispersed among the fungal hyphae. Lichens survive in a wide variety of environments, either forming small, circular crusts or leaflike structures attached to bark, rocks, or soil, or as hairlike structures hanging from tree branches.

The relationship between the fungus and the alga in a lichen is an example of mutualism, in which both partners benefit from the partnership. This relationship enables each to tolerate harsh conditions where neither could survive alone. In this partnership, the fungus furnishes the alga with water, prevents overexposure to sunlight, and provides simple mineral nutrients, while the photosynthesizing alga supplies food to the fungus even if no other organic material is available. In dry, barren areas where plants have a hard time growing, such as polar tundra, deserts, rocky outcrops, or high mountains, lichens are the primary photosynthesizers. Some remarkable species even grow inside porous rocks, just below the surface where some light can still reach the algal cells.

Their ability to grow in severe conditions often makes lichens pioneers in plant succession, the process in which plants colonize bare rock or soil. Lichens release acids that break down inhospitable rock, permitting soil-trapping mosses and grasses to grab hold. In areas where soil gradually accumulates, such as a forest floor, the pioneering lichens are eventually replaced by plants and trees, although other lichens may grow on these plants and trees.

Structure

About 18,000 to 20,000 species of lichens have been identified. Scientists typically classify lichens based on the fungal partner, and the lichen name is the same as the scientific name of the fungal partner. By far the most common fungi found in lichens are sac fungi, or ascomycetes, which produce reproductive spores in special cells shaped like sacs. In about two dozen tropical lichens, the fungal component is a club fungus, or basidiomycete.

The club fungi, which include common mushrooms and toadstools, produce their reproductive spores in special cells that are shaped like clubs. The algal partner of a lichen is usually composed of green algae in the form of single cells or chains of cells. Green algae contain chlorophyll, the primary light-absorbing pigment necessary for photosynthesis. This pigment is housed in saclike structures called chloroplasts, the sites of photosynthesis. In some dark-coloured, gelatinous lichens, cyanobacteria (also known as blue-green algae) are the algal component. Usually a lichen contains a single species of algae, but in a few cases both a green alga and a cyanobacterium live in different parts of the same lichen.

A typical lichen has a three-layered structure. A middle layer containing algal cells entwined in fungal hyphae is sandwiched between two layers of fungal tissue. This three-layered structure is arranged into one of three basic growth forms. *Crustose* lichens grow as flattened crusts with the bottom layer cemented to the surface of a rock or tree bark. *Foliose* lichens have a leaflike appearance, with a distinct upper and lower surface exposed to the air.

Foliose lichens often form large, flaky patches on tree trunks. *Fruticose* lichens grow in hairlike, shrubby strands on the ground or hanging from tree branches. Each strand is tubular, with the typical three-layered structure surrounding a hollow core. A common fruticose lichen is old man's beard, which hangs in wispy clumps from tree limbs and resembles moss.

Growth and Reproduction

Lichens lack roots, but they are usually firmly attached to the surfaces where they grow by hyphae. Lichens grow excruciatingly slowly, adding a few millimeters to their length or diameter in a year. The fastest growing varieties may add no more than 30 mm (1.2 in) to their length in a year. Lichens may have long life spans—a lichen found in West Greenland in the Arctic is more than 4,500 years old. Scientists have used the sizes of large specimens to estimate how long it has been since glaciers covered arctic and mountain areas.

During prolonged dry periods, lichens survive by retaining a small amount of water and reducing their growth and metabolic processes to the barest survival levels. After a rain, the fungal partner is able to soak up water like a sponge, absorbing two to three times its weight in water. A moist internal environment is critical for the alga, which needs water, along with carbon dioxide and sunlight, to manufacture food through photosynthesis. During dry periods when other food may be lacking, the fungus benefits by absorbing sugars and nitrogen-rich compounds produced by the alga during photosynthesis. In some cases the fungal partner is parasitic, using special fungal hyphae called haustoria to penetrate the algal cell and absorb nourishment. Since the fungal partner usually does not kill the alga, some scientists call this relationship a controlled parasitism of the alga by the fungus.

In lichen reproduction, the fungus undergoes sexual reproduction and the alga undergoes asexual reproduction, each independently of its partner. In the fungus, the hyphae contains two mating strains, called plus and minus strains because there are no anatomical distinctions between them. The nuclei from these two mating strains fuse within a special, multicellular reproductive structure on the surface of the lichen. The fused nuclei divide several times and produce spores, which are carried away by the wind and may germinate to form a new mass of hyphae. These new fungal hyphae are free of algae and typically must link with an alga in order to survive. The algal cells within the lichen reproduce through mitosis, a process in which a single cell divides into two genetically identical cells.

Lichens as a unit may also undergo asexual reproduction. The lichen may slough off small clumps of *soredia,* intertwined fungal hyphae with a few algal cells, which disperse to new habitats. Alternatively, the lichen may produce small, fingerlike buds on the surface called *isidia,* which break off and form new lichens.

Importance of Lichens

Lichens are common food for insects and slugs. In the arctic tundra, reindeer and caribou rely on lichens during the winter, when no other food is available. Several species of lichens that sprout up through the snow are called reindeer mosses.

Humans rarely eat lichens except in cases where no other food is available. The Bible may chronicle one such example: Some scholars believe that the manna or bread that the ancient Israelites ate in the desert when they fled Egypt was made of lichen. In Japan, where algae as well as fungi are prized foods, certain lichens are eaten as delicacies.

Although lichens have been used in folk medicine as purported cures for many ills, from headaches and toothaches to tuberculosis, diabetes, and asthma, their use in modern medicine is recent. The discovery in the 1940s that some fungi produce potent antibiotics stimulated an extensive screening of fungi and lichens. Since then, lichen extracts have found limited use in Europe, where lichen antibiotics have been used to treat tuberculosis and some skin diseases.

Lichen extracts are also used to add colour or scent. Many lichens have earthy red, brown, russet, and blue pigments, and these pigments have been used as clothing dyes since the time of ancient Greece. Native Americans use boiled lichen extracts to dye cloth and baskets.

Even the famous Harris tweeds, woolen textiles from the Scottish islands, are still dyed with Scottish lichens. A number of dark, oily extracts from European and African lichens are used to add scents to soaps and perfumes.

Lichens are used as living indicators of environmental problems because of their sensitivity to atmospheric pollution. Despite their hardiness in severe conditions, many lichens are damaged by the chemicals found in polluted air and by acid rain produced from the burning of coal, oil, and gasoline, and other industrial processes.

Comparisons of lichen damage from industrialized and nonindustrialized areas provide an index of air pollution. Studies spanning the past 50 years have found that the variety and number of lichens have declined in many industrialized cities in both Europe and North America.

Fortunately, studies have shown that when a pollution source is shut down, lichens can make a comeback, providing hope for their long-term survival.

CHLOROPLAST

Chloroplast, structure in the cells of plants and algae where photosynthesis takes place. Chloroplasts are mostly disk-shaped organelles, 4 to 6 micrometers in diameter. They occur most abundantly in leaf cells, where they can apparently orient themselves to light. Perhaps 40 to 50 chloroplasts exist in one cell and 500,000 in each sq mm (0.06 sq in) of leaf surface. Each chloroplast is enclosed in a double membrane.

Internally, it consists of a ground substance called the stroma, which is traversed by a complex network of interconnected disks called thylakoids. Many of the thylakoids are stacked like saucers; the stacks are called grana. Molecules of chlorophyll, which absorb light for photosynthesis, are attached to the thylakoids. The light energy captured by the chlorophyll is converted to adenosine triphosphate, or ATP, in a series of chemical reactions that take place in the grana. Chloroplasts also contain small starch grains that temporarily store the products of photosynthesis.

In plants, chloroplasts develop in the presence of light from small, colourless organelles called proplastids. As cells divide in the growing parts of a plant, the proplastids inside them divide by fission. Thus, the daughter cells have the ability to produce chloroplasts. In algae, chloroplasts divide directly, without developing from proplastids. The self-reproducing ability of chloroplasts, their bacteria-like DNA and ribosomes, and their close similarity regardless of the type of cell they inhabit, suggest that they were once independent organisms that come to exist in symbiosis with the plant cell as host.

Adenosine Triphosphate

Adenosine Triphosphate (ATP), molecule found in all living organisms that is the main immediate source of usable energy for the activities of the cells. ATP is built up by the metabolism of foodstuffs in the cell in special compartments called mitochondria. Because the energy-exchanging function of ATP and the catalytic (work-boosting) function of enzymes are intimately connected, ATP is characterized as a coenzyme.

The adenosine part of the molecule is made up of adenine, a nitrogen-containing compound (also one of the principal components of the gene), and ribose, a five-carbon sugar. Three phosphate units (triphosphate), each made up of one phosphorus atom and four oxygen atoms, are attached to the ribose. The two bonds between the three phosphate groups are high-energy bonds, that is, they are relatively weak and yield their energy readily when split by enzymes. With the release of the end phosphate group, 7 kilocalories (7 calories, in common usage) of energy become available for work, and the ATP molecule becomes ADP (adenosine diphosphate).

Most of the energy-consuming reactions in cells are powered by the conversion of ATP to ADP; they include the transmission of nerve signals, the movement of muscles, the synthesis of protein, and cell division. Usually, ADP quickly regains the third phosphate unit through the action of cytochrome, a protein that builds it up by using food energy. In vertebrate muscle and brain cells, excess ATP can join with creatine to provide a reserve energy store.

The release of two phosphate groups from ATP by the enzyme adenyl cyclase forms AMP (adenosine monophosphate), a nucleotide component of nucleic acids, the material of DNA; this enzyme is important in many of the body's reactions. The American biochemist Earl Sutherland, Jr., won the 1971 Nobel Prize in physiology or medicine for work showing that one form of AMP called cyclic AMP, created by the action of the enzyme adenyl cyclase, is instrumental in the activities of many hormones, including epinephrine and ACTH.

Epinephrine

Epinephrine, hormone secreted by the adrenal gland. The pure compound, first isolated by the Japanese chemist Jokichi Takamine, is also known as adrenaline. It was formerly prepared from adrenal glands, but it is now made synthetically.

Epinephrine normally is present in the bloodstream in minute quantities. In times of excitement or emotional stress, however, large additional quantities are secreted, exerting a marked effect on body structures in preparation for physical exertion. Epinephrine stimulates the heart, constricts the small blood vessels, raises the blood pressure, liberates sugar stored in the liver, and relaxes certain involuntary muscles while contracting others. It is widely used as a drug to stimulate the heart in cases of shock, to prevent bleeding, and to expand lung bronchioles in acute asthma attacks.

The adrenal gland also secretes a substance chemically related to epinephrine, called norepinephrine, noradrenaline, or levoarterenol. In general, the function of norepinephrine seems to be the maintenance of normal blood circulation.

It is also the chemical agent responsible for transmission of nerve impulses in the autonomic nervous system. Large amounts of epinephrine and norepinephrine are produced by some tumors of the adrenal glands, resulting in a great increase in blood pressure.

Shock

Shock, in physiology, state of acute circulatory insufficiency of the blood. It is the result of the inability of the heart to pump an adequate volume of blood at sufficient pressure for normal blood passage through the major organs of the body. Shock may be caused by heart failure, injury, burns, hemorrhage, or major surgery.

Sudden infection or poisoning and abnormal extracellular fluid volume resulting from excessive loss of water and electrolytes (ionic conductors) from the gastrointestinal tract, kidneys, or skin may also cause shock.

Shock is generally characterized by apathy, weakness, shallow breathing, rapid heartbeat, feeble pulse, lowered blood pressure, and coldness and clamminess of skin. During the early stage consciousness is retained, but alertness is diminished. Sudden peripheral circulatory failure, however, affects the brain and causes fainting. In less severe shock compensatory constriction of the blood vessels helps restore circulation, but if shock persists, compensatory mechanisms fail and local anemia damages vital organs, such as the brain, heart, liver, and lungs.

First aid for shock includes having the victim lie down, keeping him or her warm but not overheated, stopping any bleeding, and, if the person is not breathing, administering artificial respiration. A physician or paramedic may administer oxygen and sedative drugs and take measures to lower the body temperature if it is very high. Medical personnel may give intravenous glucose and salt solutions, and transfusions of plasma or whole blood may be required when shock is due to blood loss. In medical centres, treatment is monitored by direct measurements of arterial blood pressure and arterial blood gases. Sometimes blood pressure is increased by vasopressors (drugs that constrict blood vessels).

AUTONOMIC NERVOUS SYSTEM

Autonomic Nervous System, in vertebrate anatomy, one of the two main divisions of the nervous system, supplying impulses to the body's heart muscles, smooth muscles, and glands. The autonomic system controls the action of the glands; the functions of the respiratory, circulatory, digestive, and urogenital systems; and the involuntary muscles in these systems and in the skin. Controlled by nerve centres in the lower part of the brain, the system also has a reciprocal effect on the internal secretions, being controlled to some degree by the hormones and exercising some control, in turn, on hormone production.

Two antagonistic divisions make up the autonomic nervous system: the sympathetic, or thoracicolumbar, division, which stimulates the heart, dilates the bronchi, contracts the arteries, and inhibits the digestive system, preparing the organism for physical action; and the parasympathetic, or craniosacral, division, which has the opposite effects, and prepares the organism for feeding, digestion, and rest.

The sympathetic division consists of a chain of interconnected ganglia (groups of nerve cells) on each side of the vertebral column, which send nerve fibres to several large ganglia, such as the coeliac ganglion. They, in turn, give rise to nerves passing to the internal organs. The ganglia of the sympathetic chains are connected to the central nervous system by fine branches connecting each ganglion with the spinal cord. Fibres of the parasympathetic system arise in the brain and, with the cranial nerves, especially the vagus and accessory nerves, pass to ganglia and plexuses (networks of nerves) within the various organs. The lower part of the body is innervated by fibres arising from the lowest (sacral) segment of the spinal cord and passing to the pelvic ganglion, which gives rise to nerves for such organs as the rectum, bladder, and genital organs.

Adrenal Gland

Adrenal Gland, vital endocrine gland that secretes hormones into the bloodstream, situated, in humans, on top of the upper end of each kidney. The two parts of the gland—the inner portion, or medulla, and the outer portion, or cortex—are like separate organs: They are composed of different types of tissue and perform different functions. The adrenal medulla, composed of chromaffin cells, secretes the hormone epinephrine, also called adrenaline, in response to stimulation of the sympathetic nervous system at times of stress. The medulla also secretes the hormone norepinephrine, which plays a role in maintaining normal blood circulation. The hormones of the medulla are called catecholamines. Unlike the adrenal cortex, the medulla can be removed without endangering the life of an individual.

The adrenal outer layer, or cortex, secretes about 30 steroid hormones, but only a few are secreted in significant amounts. Aldosterone, one of the most important hormones, regulates the balance of salt and water in the body. Cortisone and hydrocortisone are necessary to regulate fat, carbohydrate, and protein metabolism.

Adrenal sex steroids have a minor influence on the reproductive system. Modified steroids, now produced synthetically, are superior to naturally secreted steroids for treatment of Addison's disease and other disorders.

Adrenocorticotropic Hormone

Adrenocorticotropic Hormone (ACTH), also known as corticotropin, hormone secreted by the anterior part of the pituitary gland. The specific function of ACTH is to stimulate the growth and secretions of the *cortex* (outer layer) of the adrenal gland. One of these secretions is cortisone, a hormone involved in carbohydrate and protein metabolism.

ACTH is a complex protein molecule containing 39 amino acids. Its molecular weight is approximately 5,000. The biological activity of the ACTH of various animal species resembles that of humans, but the sequence of amino acids varies somewhat between species.

ACTH production is controlled in part by the hypothalamus and in part by the existing levels of adrenal gland hormones.

ACTH levels increase in response to stress, disease, and decreased blood pressure. Associated medical conditions include Addison's disease and Cushing's syndrome.

PITUITARY GLAND

Pituitary Gland, master endocrine gland in vertebrate animals. The hormones secreted by the pituitary stimulate and control the functioning of almost all the other endocrine glands in the body. Pituitary hormones also promote growth and control the water balance of the body.

The pituitary is a small bean-shaped, reddish-gray organ located in the saddle-shaped depression (sella turcica) in the floor of the skull (the sphenoid bone) and attached to the base of the brain by a stalk; it is located near the hypothalamus.

The pituitary has two lobes—the anterior lobe, or adenohypophysis, and the posterior lobe, or neurohypophysis—which differ in structure and function. The anterior lobe is derived embryologically from the roof of the pharynx and is composed of groups of epithelial cells separated by blood channels; the posterior lobe is derived from the base of the brain and is composed of nervous connective tissue and nervelike secreting cells. The area between the anterior and posterior lobes of the pituitary is called the intermediate lobe; it has the same embryological origin as the anterior lobe.

The Anterior Lobe

Concentrated chemical substances, or hormones, which control 10 to 12 functions in the body, have been obtained as extracts from the anterior pituitary glands of cattle, sheep, and swine. Eight hormones have been isolated, purified, and identified; all of them are peptides, that is, they are composed of amino acids. Growth hormone (GH), or the somatotropic hormone (STH), is essential for normal skeletal growth and is neutralized during adolescence by the gonadal sex hormones.

Thyroid-stimulating hormone (TSH) controls the normal functioning of the thyroid gland; and the adrenocorticotropic hormone (ACTH) controls the activity of the cortex of the adrenal glands and takes part in the stress reaction. Prolactin, also called lactogenic, luteotropic, or mammotropic hormone, initiates milk secretion in the mammary gland after the mammary tissues have been prepared during pregnancy by the secretion of other pituitary and sex hormones.

The two gonadotropic hormones are follicle-stimulating hormone (FSH) and luteinizing hormone (LH). Follicle-stimulating hormone stimulates the formation of the Graafian follicle in the female ovary and the development of spermatozoa in the male. The luteinizing hormone stimulates the formation of ovarian hormones after ovulation and initiates lactation in the female; in the male, it stimulates the tissues of the testes to elaborate testosterone. In 1975 scientists identified the pituitary peptide endorphin, which acts in experimental animals as a natural pain reliever in times of stress.

Endorphin and ACTH are made as parts of a single large protein, which subsequently splits. This may be the body's mechanism for coordinating the physiological activities of two stress-induced hormones. The same large prohormone that contains ACTH and endorphin also contains short peptides called melanocyte-stimulating hormones. These substances are analogous to the hormone that regulates pigmentation in fish and amphibians, but in humans they have no known function.

Research has shown that the hormonal activity of the anterior lobe is controlled by chemical messengers sent from the hypothalamus through tiny blood vessels to the anterior lobe. In the 1950s, the British neurologist Geoffrey Harris discovered that cutting the blood supply from the hypothalamus to the pituitary impaired the function of the pituitary.

In 1964, chemical agents called releasing factors were found in the hypothalamus; these substances, it was learned, affect the secretion of growth hormone, a thyroid-stimulating hormone called thyrotropin, and the gonadotropic hormones involving the testes and ovaries. In 1969 the American endocrinologist Roger Guillemin and colleagues isolated and characterized thyrotropin-releasing factor, which stimulates the secretion of thyroid-stimulating hormone from the pituitary.

In the next few years his group and that of the American physiologist Andrew Victor Schally isolated the luteinizing hormone-releasing factor, which stimulates secretion of both LH and FSH, and somatostatin, which inhibits release of growth hormone. For this work, which proved that the brain and the endocrine system are linked, they shared the Nobel Prize in physiology or medicine in 1977. Human somatostatin was one of the first substances to be grown in bacteria by recombinant DNA.

The presence of the releasing factors in the hypothalamus helped to explain the action of the female sex hormones, estrogen and progesterone, and their synthetic versions contained in oral contraceptives, or birth-control pills. During a woman's normal monthly cycle, several hormonal changes are needed for the ovary to produce an egg cell for possible fertilization. When the estrogen level in the body declines, the follicle-releasing factor (FRF) flows to the pituitary and stimulates the secretion of the follicle-stimulating hormone.

Through a similar feedback principle, the declining level of progesterone causes a release of luteal-releasing factor (LRF), which stimulates secretion of the luteinizing hormone. The ripening follicle in the ovary then produces estrogen, and the high level of that hormone influences the hypothalamus to shut down temporarily the production of FSH. Increased progesterone feedback to the hypothalamus shuts down LH production by the pituitary.

The daily doses of synthetic estrogen and progesterone in oral contraceptives, or injections of the actual hormones, inhibit the normal reproductive activity of the ovaries by mimicking the effect of these hormones on the hypothalamus.

The Intermediate Lobe

In lower vertebrates this part of the pituitary secretes

melanocyte-stimulating hormone, which brings about skin-colour changes. In humans, it is present only for a short time early in life and during pregnancy, and is not known to have any function.

The Posterior Lobe

Two hormones are secreted by the posterior lobe. One of these is the antidiuretic hormone (ADH), vasopressin. Vasopressin stimulates the kidney tubules to absorb water from the filtered plasma that passes through the kidneys and thus controls the amount of urine secreted by the kidneys. The other posterior pituitary hormone is oxytocin, which causes the contraction of the smooth muscles in the uterus, intestines, and blood arterioles.

Oxytocin stimulates the contractions of the uterine muscles during the final stage of pregnancy to stimulate the expulsion of the fetus, and it also stimulates the ejection, or let-down, of milk from the mammary gland following pregnancy. Synthesized in 1953, oxytocin was the first pituitary hormone to be produced artificially. Vasopressin was synthesized in 1956.

Pituitary Disturbances

Pituitary functioning may be disturbed by such conditions as tumors, blood poisoning, blood clots, and certain infectious diseases. Conditions resulting from a decrease in anterior-lobe secretion include dwarfism, acromicria, Simmonds's disease, and Fröhlich's syndrome.

Dwarfism occurs when anterior pituitary deficiencies occur during childhood; acromicria, in which the bones of the extremities are small and delicate, results when the deficiency occurs after puberty.

Simmonds's disease, which is caused by extensive damage to the anterior pituitary, is characterized by premature aging, loss of hair and teeth, anemia, and emaciation; it can be fatal. Fröhlich's syndrome, also called adiposogenital dystrophy, is caused by both anterior pituitary deficiency and a lesion of the posterior lobe or hypothalamus.

The result is obesity, dwarfism, and retarded sexual

development. Glands under the influence of anterior pituitary hormones are also affected by anterior pituitary deficiency.

ADRENAL GLAND

Adrenal Gland, vital endocrine gland that secretes hormones into the bloodstream, situated, in humans, on top of the upper end of each kidney. The two parts of the gland—the inner portion, or medulla, and the outer portion, or cortex—are like separate organs: They are composed of different types of tissue and perform different functions.

The adrenal medulla, composed of chromaffin cells, secretes the hormone epinephrine, also called adrenaline, in response to stimulation of the sympathetic nervous system at times of stress. The medulla also secretes the hormone norepinephrine, which plays a role in maintaining normal blood circulation. The hormones of the medulla are called catecholamines. Unlike the adrenal cortex, the medulla can be removed without endangering the life of an individual.

The adrenal outer layer, or cortex, secretes about 30 steroid hormones, but only a few are secreted in significant amounts. Aldosterone, one of the most important hormones, regulates the balance of salt and water in the body. Cortisone and hydrocortisone are necessary to regulate fat, carbohydrate, and protein metabolism. Adrenal sex steroids have a minor influence on the reproductive system. Modified steroids, now produced synthetically, are superior to naturally secreted steroids for treatment of Addison's disease and other disorders

HYDROCORTISONE

Hydrocortisone, also cortisol, common names for 17-hydroxy-corticosterone, the principal hormone secreted by the outer layer, or cortex, of the adrenal gland.

Hydrocortisone affects the metabolism of carbohydrate, protein, and fat; the maturation of white blood cells; the retention of salt and water in the body; the activity of the nervous system; and the regulation of blood pressure. Secretion of hydrocortisone from the adrenal cortex is stimulated by the pituitary hormone ACTH.

Because of their widespread effects, hydrocortisone and related compounds, called corticosteroids, or corticoids, are employed for many medical purposes. They are used to treat a deficiency of adrenal cortical hormones, a condition called congenital adrenal hyperplasia, rheumatoid disease that is not helped by milder drugs, and to counter severe non-infectious inflammations. Corticosteroids suppress the immune response, so they are used to increase acceptance of transplants (Transplantation, Medical).

Other conditions for which they are helpful are asthma, collagen diseases, and eye inflammations. Because corticosteroids affect so many body processes, they must be used carefully. Corticosteroids dispose persons to infection and can lead to swelling of the face and limbs, muscle weakness, weight gain, high blood pressure, and diabetes.

Another naturally occurring corticosteroid hormone, called cortisone, was the first corticosteroid to be isolated, in 1935. It was synthesized in 1944 and subsequently became available for widespread medical use. Cortisone is rapidly converted to hydrocortisone in the body. Synthetic corticosteroids with more specific activity have now been made and are preferred in many situations.

Chapter 5

Oomycetes

The Oomycetes are the largest group of heterotrophic Stramenopiles. They are found all over the world in fresh and salt water habitats. Some of the terrestrial Oomycetes are among the most important plant pathogenic organisms that may be facultatively or obligately parasitic. The organisms range from those preferring well-aerated streams to some that occur in stagnant waters and that may gain most of their energy from anaerobic metabolism.

Most members of the group are filamentous and lack septa except where reproductive cells are produced, but holocarpic forms are found in the Saprolegniales and Lagenidiales. The cell walls of the group consist mostly of ß-1,3- and ß-1,6-glucans with a small amount of cellulose. A few members also have chitin deposits.

All members of the group that have had extensive life cycle studies have been found to be diploid in the vegetative phase with meiosis occurring during gametogenesis. The group is characterized by oogamous reproduction that includes interaction of male antheridia with female oogoni

HAPTOGLOSSA

Haptoglossa is an enigmatic protistan genus provisionally placed in the phylulm Oomycota (Order Saprolegniales, Family Ectrogellaceae). Consisting of about seven species, members of this unique taxon are commonly found in manure piles and herbivore dung as obligate parasites of bactivorus nematodes and rotifers. The most commonly infected nematode hosts include the genera, *Rhabditis* and *Bunonema*.

Some species produce motile biflagellate zoospores with smooth flagella as a part of their life cycle while others have been described as being "aplanosporic" meaning that they do not produce zoospores. Their morphology like many other protists is simple.

The thalli are oblong, following the morphology of the host, and holocarpic in form. Discharge tubes become prominent at maturity and rupture through the host cuticle in order to release zoospores into the environment. Once the zoospores are released, they swim around for a short amount of time before settling down and encysting. The cyst then develops into what is called a "gun cell".

The uninucleate gun cell when brushed by a grazing rotifer or nematode, fires and injects its protoplast called the "sporidium" into the host. The sporidium will then develop into a single holocarpic thallus. So far no one has discovered a sexual stage (no oospores) or any kind of resting spore for any of the species, leaving the life cycle incomplete. Characters which have been used to separate out the species include: spore morphology and structure (Aplanosporic species) and variations in gun cell structure (Zoosporic species). According to Beakes and Glockling, due to their overall cell structure and zoospore ultrastructure, members of Haptoglossa belong within the Stramenopile group but may represent a separate class based on their lack of oogamy and the unique infection cells.

FUNGUS

Fungus, any member of a diverse group of organisms that—unlike plants and animals—obtain food by absorbing nutrients from an external source. The fossil record suggests that fungi were present 550 million years ago and may have evolved even earlier. Today thousands of different types of fungi grow on and absorb food from substances such as soil, wood, decaying organic matter, or living plants and other organisms.

They range from tiny, single-celled organisms invisible to the naked eye to some of the largest living multicellular organisms. In Michigan the underground portion of an individual *Armillaria* mushroom, a type of fungus, extends more than 12 hectares (30 acres). Other fungi are among the longest-lived organisms on Earth—some lichens, a living partnership of a fungus and an alga, are thought to be more than 4,500 years old.

A large and widely distributed group of organisms, fungi perform activities essential to the functioning of all natural ecosystems. They are among the foremost decomposers of organic matter, breaking down plant and animal remains and wastes into their chemical components. As such, fungi play a critical role in the recycling of minerals and carbon. Fungi's value to humankind is inestimable.

Certain types of fungi, including several types of mold, have proven extremely valuable in the synthesis of antibiotics and hormones used in medicine and of enzymes used in certain manufacturing processes. Some fungi, such as mushrooms and truffles, are considered tasty delicacies that enhance a wide variety of recipes. Not all fungi are beneficial—some damage agricultural crops, cause disease in animals and humans, and form poisonous toxins in food.

Common fungi include mushrooms, puffballs, truffles, yeasts, and most mildews, as well as various plant and animal *pathogens* (disease agents), such as plant rusts and smuts. Some experts estimate that there are 1.5 million fungus species, of which approximately 100,000 have been identified. The unique characteristics of fungi led scientists to classify these important organisms into a separate kingdom, Kingdom Fungi (also known as Mycetae). Certain fungus-like organisms, such as downy mildews, water molds (also known as oomycetes), and slime molds, once classified as fungi, are now placed in the Kingdom Protista.

Unique Feeding Method

Fungi lack chlorophyll, the green pigment that enables plants to make their own food. Consequently, fungi cannot synthesize their own food the way plants do. In order to feed, fungi release digestive enzymes that break down food outside their bodies. The fungus then absorbs the dissolved food through its cell walls.

Depending as they do on outside sources for food, fungi have developed various living arrangements that enhance their opportunities for food absorption. Some fungi live as parasites, feeding on living plants, animals, and even other fungi. Certain fungus parasites injure plants and animals, causing millions of dollars of damage to farm animals, crops, and trees each year. The fungus *Ophiostoma ulmi*, which causes Dutch elm disease, has killed tens of millions of elm trees around the world.

Fungi that obtain their food by breaking down dead organisms or substances that contain organic compounds, such as starch and cellulose, are called saprobes or saprophytes. While they are invaluable decomposers of organic material, saprobes can also cause food spoilage and destroy wood products. During the American Revolution (1775-1783), more British ships were destroyed by wood-digesting saprobes than by enemy attack. Some saprobes even grow in aviation fuels, where they breakdown the fuels, destroying their usefulness.

Some fungi also form highly specialized relationships with other organisms. The roots of most plants develop a mutually beneficial association with fungi to form mycorrhizae. Mycorrhizae greatly increase the nutrient-absorbing capacity of the plant root—the fungus absorbs minerals from the soil and exchanges them for organic nutrients synthesized by the plant. Fungi also form mutualistic associations with various animals. Leaf-cutting ants cut pieces of leaves and bring them into their underground nests, where they feed them to certain fungi.

These fungi primarily live in ant nests, and the ants eat nothing but the fungi. Some termites and wood-boring beetles use fungi to break down the cellulose in wood, making the wood easier for the insects to digest. Plant parasites such as rusts invade plant cells via specialized structures called haustoria that absorb nutrients from the cell.

Fungi Structure

With the exception of one-celled species, most fungi are composed of threadlike tubular filaments called hyphae. Each individual hypha is surrounded by a fairly rigid wall usually made of chitin—the same material that forms the exoskeletons of insects. Hyphae that are partitioned by dividing cross walls are called septate hyphae, and hyphae without cross walls are called nonseptate, or coenocytic, hyphae. Fungal cells contain cytoplasm, which is a mixture of internal fluids and nutrients. Cytoplasm flows freely within the hyphae, providing nutrients wherever they are needed.

Hyphae grow by elongation at the tips and by branching to form an interwoven mat known as the mycelium. As the mycelium develops, it may produce large fruiting bodies or other structures that contain reproductive spores. Fruiting bodies are often the most visible structure of a fungus, usually growing above the soil or other surfaces so that the spores can be dispersed by air currents or other mechanisms. In contrast, the mycelium is usually hidden beneath the surface of the plant, animal, or other material it is decomposing. A mushroom mycelium is typically buried beneath the soil surface, while its fruiting body, the familiar umbrella-shaped structure, sprouts from the ground.

Reproduction

The wide variety of fungi demonstrate many reproductive methods. In general, most fungi reproduce by making tiny spores. Fungi typically produce large numbers of spores. A giant puffball, produces an estimated 7 trillion spores.

Fungi typically follow a reproductive cycle that involves the production of sexual spores. These spores contain one or more nuclei and are usually haploid—that is, their nuclei contain one set of chromosomes. When environmental conditions are favorable, the spores germinate and develop into a mycelium that produces fruiting bodies with enormous numbers of sexual spores, which repeat the reproductive cycle. Some fungi produce asexual spores directly from hyphae, which then germinate to produce additional mycelium. The mycelium spreads rapidly, aiding the fungus in dispersal and colonization.

In the reproductive cycle of mushrooms, the mycelium contain hyphae of two mating types, sometimes called plus and minus strains, with no obvious anatomical differences distinguishing them. If plus and minus strains of hyphae fuse, sexual reproduction begins.

Initially the nuclei of the two hyphae remain separate, producing an intermediate stage called the *dikaryon,* meaning "two nuclei." The dikaryon stage can last from weeks to years, depending upon the species. The two nuclei in the dikaryon eventually fuse to produce a diploid cell—that is, a cell that contains one nucleus with two sets of chromosomes.

This cell immediately undergoes meiosis, a type of nuclear cell division that produces offspring with half the genetic material as the parents. Meiosis usually produces four genetically unique haploid spores and the reproductive cycle begins again.

This population of genetically different spores has a better chance of surviving environmental changes, such as disease or temperature changes, that may wipe out an entire population of genetically identical spores.

Many fungi can reproduce by the fragmentation of their hyphae. Each fragment develops into a new individual. Yeast, a small, single-celled fungus, reproduces by budding, in which a bump forms on the yeast cell, eventually partitioning from the cell and growing into a new yeast cell.

Classification of Fungi

Scientists have long disagreed about how to classify fungi, and the classification systems are still developing. The first description of fungi by Italian botanist Pier Antonio Micheli. Fungi were initially classified in the Plant Kingdom, and the field of fungus study, or mycology, developed as a branch of botany. Recognition of the unique characteristics of fungi led mycologists to establish a separate kingdom, Kingdom Fungi, in the late 1960s. More recently, some mycologists have noted that some organisms, such as slime molds, downy mildews, and water molds, have characteristics that place them in the Kingdom Protista rather than the fungi. Unlike true fungi, some slime molds have a mobile, multinucleate feeding stage similar to amoebas.

Downy mildews and water molds produce motile cells for part of their life cycle, have hyphal walls that lack chitin, and make an egg cell and sperm nuclei. Some scientists have proposed that downy mildews and water molds deserve to be classified in a separate kingdom, called Kingdom Stramenopila.

Fungi are classified primarily by the type of spores and fruiting bodies produced. Many mycologists divide the Kingdom Fungi into four main phyla: Chytridiomycota, Zygomycota, Ascomycota, and Basidiomycota. A fifth phylum, Deuteromycota, is used by some taxonomists for fungi that typically produce only asexual spores.

The phylum Chytridiomycota, commonly called Chytrids, includes approximately 800 species that are found in aquatic (freshwater and marine) or moist habitats. Chytrids are among the smallest and simplest fungi. Most have a central body with small tubelike extensions, while others produce a small network of hyphae.

Chytrids develop a structure called a sporangium that has motile spores equipped with a posterior flagellum, a long, whiplike tail that aids in locomotion. Chytrids grow as saprobes in damp soils and water, or as parasites of plants, animals, algae, protists, and other fungi. Some do not require oxygen and live only in the guts of plant-eating animals, where they break down material containing cellulose and other compounds. Because chytrid spores are motile, some mycologists have classified them in the Kingdom Protista.

The Zygomycota include approximately 900 terrestrial species, including many important decomposers, mycorrhizal fungi, and parasites of spiders and insects. One of the most common zygomycetes is black bread mold, often found on bread, fruit, and other foods. The fungus looks like a fuzzy growth with tiny black dots at the tips of the fuzz. The black dots are sporangia growing at the ends of special hyphae. The sporangia produce asexual, nonswimming spores called sporangiospores. Zygomycetes reproduce sexually by forming thick-walled zygospores.

The largest group of fungi, with around 50,000 known species, is the Ascomycota, or sac fungi. This group includes yeasts, lichens, morels, cup fungi, truffles, and a number of plant parasites such as powdery mildews. Named for the sexual spores produced inside saclike cells called *asci* (singular, *ascus)*, Ascomycota also may produce very fine, almost powdery asexual spores called *conidia*. Certain Ascomycota such as cup fungi produce fruiting bodies with sexual spores on their upper surface, while others, including the truffles, produce sexual spores inside tuber-like fruiting bodies that develop underground.

Ascomycetes are used to produce Camembert and Roquefort cheeses. The slight grittiness in these cheeses is due to conidia being crushed between the teeth. The mold ergot, which infects the flowers of rye and other grains, produces toxins that can poison humans and other animals that eat the infected grain. The yeast *Candida albicans* is a common pathogen of humans, causing such ailments as oral thrush and vaginal yeast infections. In people with weakened immune systems, this yeast may spread widely throughout the body and become life threatening.

The Basidiomycota, also known as club fungi, include around 25,000 species of mushrooms, puffballs, bird's nest fungi, jelly fungi, rusts, smuts, and shelf and bracket fungi. This division contains important plant parasites, mutualists, and saprobes, including decay fungi that cause brown rot and white rot of wood. These fungi are named for their specialized, club-shaped reproductive cells, called basidia, which form spores called basidiospores. Basidia may line gills or tubes on the underside of fleshy fruiting bodies, which consists of a stalk and cap—the familiar components of most mushrooms. Certain Basidiomycota produce spores inside tuber-like underground fruiting bodies, called "false truffles."

Many basidiomycetes are saprobes, which play a vital role in the decomposition of litter, wood, and dung. A number of mushrooms are good to eat, such as boletes and chantarelles, both of which are highly prized for their distinct flavour. Other mushrooms are well known for their poisonous qualities, including the death cap (*Amanita phalloides*). Some, such as the liberty cap (*Psilocybe semilanceata*) and the fly agaric (*Amanita muscaria*), are well known for their hallucinogenic properties. Smuts—such as *Ustilago*, which attacks corn, and stinking smut (*Tilletia*), which attacks wheat—are common basidiomycetes that invade flowering plants, especially cereal grasses, and cause serious economic loss. Rusts, such as *Puccinia*, which attacks wheat, invade plant cells of agricultural crops and forest trees, causing millions of dollars in losses each year.

The Deuteromycota, or imperfect fungi, comprise about 25,000 species, many of which do not have a defined sexual cycle. They typically reproduce asexually by spores called conidia on specialized hyphae called conidiophores. The deuteromycetes include many molds, some of which are important to humans. *Penicillium*, the mold used to develop the first antibiotic, is sometimes classified in the Deuteromycota. On the other side of the ledger, the deuteromycetes also include organisms such as ringworm that are serious animal and plant pathogens.

Uses of Fungi

Fungi have been used as a food source since the beginning of recorded history. Mushrooms add flavour, texture, and nutritional value to many dishes. In North America in recent years, a variety of mushrooms have gained popularity, including portabella, cremini, oyster, morel, chantarelle, wood or tree ear, truffle, matsutake, and shiitake.

Truffles—tuber-like, fleshy fungi with a characteristic taste and aroma—are highly prized by gourmet chefs. Harvested most commonly in France and northern Italy, truffles are collected with the aid of trained dogs or pigs that use scent to hunt these fungi hidden beneath the soil. The price for truffles in Europe may reach as high as $500 (U.S.) per pound in some years.

Other fungi are used in the manufacture of foods. Yeast, is added to fruit juice, which it ferments to produce wine. Yeasts also are used in the manufacturing of beer, and they are added to dough to make bread rise, producing more volume and a lighter texture in the final baked product. Certain molds are used to ripen cheeses, such as Brie, Camembert, and the characteristic blue-veined Roquefort. In Asia, fungi are added to soybeans and allowed to ferment to make several food products—soy sauce is made with the mold *Aspergillus*, and tempeh is made with the black bread mold *Rhizopus*.

Many fungi also produce biologically active compounds that are useful in manufacturing. These compounds include alcohols—such as ethanol and glycerol produced during fermentation—and plant growth regulators—such as giberellic acid, which is used in the promotion of plant and fruit development. Fungi are extremely important in the production of antibiotics; penicillin, griseofulvin, cyclosporine, and cephalosporin are used to fight bacterial and fungal diseases worldwide.

Fungi are becoming an increasingly important tool in cleaning the environment. The accumulation of pesticides and other chemicals in the environment is destroying many ecosystems, and placing many animal and plant species at risk. A number of fungi are used in bioremediation, in which the fungi are mixed with polluted water or soil, where they decompose the organic material in pollutants and, in the process, detoxify them. Fungi employed in this effort include many that are commonly found in soils, such as *Aspergillus, Fusarium, Rhizopus, Mucor, Penicillium,* and *Trichoderma*. Fungi also have been used successfully to control insects, fungus pathogens, roundworms, and other organisms that cause damage and disease to agricultural crops.

Harmful Fungi

Fungi cause about 100,000 diseases of plants, including about 70 percent of the major crop diseases, resulting in an economic loss of billions of dollars each year. These plant pathogens cause extensive disease to seeds, seedlings, mature plants, and aging plants, resulting in decreased growth and reproduction of crop plants. Fungi also attack forest trees and wooden structures.

A number of fungi cause diseases in humans and other vertebrates. In general, these fungus infections, or mycoses, develop slowly, recur more frequently than bacterial infections, and do not produce a lasting immunity in the body. A mycosis is classified in one of two groups, depending on the part of the body that is infected.

A dermatomycosis is an infection of the skin, hair, or nails, such as ringworm or athlete's foot. These infections rarely progress to the internal organs. Most respond well to medication, although treatment may take several weeks.

A systemic mycosis, which is an infection of the entire body, is typically more serious and can be fatal for individuals whose immune system has been weakened by diseases such as acquired immunodeficiency syndrome (AIDS) or cancer. Fungal infections are typically spread by spores that enter the body through inhalation or through an opening in the skin. Some infections are passed from animals to humans or between humans. A few drugs are effective at treating systemic infections, but because treatment may last for several months to years to prevent relapse of the infection, these drugs often cause toxic side effects.

Fungi cause a number of human respiratory diseases. Coccidioidomycosis is caused by the yeast *Coccidioides immitis*. Although typically contracted by the inhalation of dust containing yeast spores, the fungus may also be introduced through the skin from infected soil. Initial symptoms may resemble the flu, with fever, cough, and possibly a skin rash, and the infection usually runs its course without specific treatment. In rare cases, the fungus penetrates internal tissues, such as the bones, joints, and brain, producing tumors that later form abscesses or ulcers. No treatment is available that can halt the course of this form of the disease.

Histoplasmosis is caused by the yeastlike fungus *Histoplasma capsulatum*, which grows in pigeon, bat, and chicken droppings. Contracted by the inhalation of dust from animal droppings, by ingestion, or through the skin. The fungus causing histoplasmosis lives as a parasite in certain tissue and blood cells of the infected person. An infection in the respiratory system is similar to tuberculosis—small spots form in the lungs—although these lesions heal on their own. A progressive form typically invades the bone marrow and is rapidly fatal.

Aspergillosis is an infection of the skin, nasal sinuses, and lungs or other internal organs caused by molds of the genus *Aspergillus*. The disease, contracted by the inhalation of spores, occurs most often among agricultural workers. Itching and pain are frequent symptoms, and if scratching is prolonged, the skin may thicken and become gray or black. A virulent type of pneumonia is caused by the yeastlike fungus *Pneumocystis carinii*, particularly prevalent in people with compromised immune systems, such as AIDS patients.

Mycotoxins are poisons produced by fungal growth in cereals, nuts, fruits, and vegetables. More than 100 species of fungi produce these toxins. The most common mycotoxin is aflatoxin, produced by *Aspergillus flavus* and *Aspergillus parasiticus*. Commonly found on corn, peanuts, and tree nuts, the toxin also can be transmitted to humans through the milk, meat, or eggs of animals fed contaminated grains. Aflatoxin is the most potent carcinogen, or potentially cancer-causing agent, yet discovered. Other mycotoxins include trichothecenes and zearalenone, compounds known to injure the intestines, bone marrow, lymph nodes, spleen, and thymus. They are produced by species of *Fusarium* that grow on grain, straw, or hay stored while damp.

Occasionally, circumstances prevent the harvesting of grains during the autumn, and the grains lie dormant in the damp fields until they are harvested in the spring. These grains are especially vulnerable to trichothecenes and zearalenone contamination. A large outbreak of trichothecenes contamination occurred in Russia in early 1944 among hungry peasants who had been searching the winter fields for unharvested wheat and millet.

ANTIBIOTICS

Antibiotics (Greek *anti*, "against"; *bios*, "life") are chemical compounds used to kill or inhibit the growth of infectious organisms. Originally the term *antibiotic* referred only to organic compounds, produced by *bacteria* or *molds*, that are toxic to other microorganisms. The term is now used loosely to include synthetic and semisynthetic organic compounds. *Antibiotic* refers generally to antibacterials; however, because the term is loosely defined, it is preferable to specify compounds as being antimalarials, antivirals, or antiprotozoals. All antibiotics share the property of selective toxicity:

They are more toxic to an invading organism than they are to an animal or human host. Penicillin is the most well-known antibiotic and has been used to fight many infectious diseases, including syphilis, gonorrhea, tetanus, and scarlet fever. Another antibiotic, streptomycin, has been used to combat tuberculosis.

Classification

Antibiotics can be classified in several ways. The most common method classifies them according to their action against the infecting organism. Some antibiotics attack the cell wall; some disrupt the cell membrane; and the majority inhibit the synthesis of nucleic acids and proteins, the polymers that make up the bacterial cell. Another method classifies antibiotics according to which bacterial strains they affect: staphylococcus, streptococcus, or *Escherichia coli*. Antibiotics are also classified on the basis of chemical structure, as penicillins, cephalosporins, aminoglycosides, tetracyclines, macrolides, or sulfonamides, among others.

Mechanisms of Action

Most antibiotics act by selectively interfering with the synthesis of one of the large-molecule constituents of the cell—the cell wall or proteins or nucleic acids. Some, however, act by disrupting the cell membrane. Some important and clinically useful drugs interfere with the synthesis of peptidoglycan, the most important component of the cell wall. These drugs include the Â-lactam antibiotics, which are classified according to chemical structure into penicillins, cephalosporins, and carbapenems. All these antibiotics contain a Â-lactam ring as a critical part of their chemical structure, and they inhibit synthesis of peptidoglycan, an essential part of the cell wall.

They do not interfere with the synthesis of other intracellular components. The continuing buildup of materials inside the cell exerts ever greater pressure on the membrane, which is no longer properly supported by peptidoglycan. The membrane gives way, the cell contents leak out, and the bacterium dies. These antibiotics do not affect human cells because human cells do not have cell walls.

Many antibiotics operate by inhibiting the synthesis of various intracellular bacterial molecules, including DNA, RNA, ribosomes, and proteins.

The synthetic sulfonamides are among the antibiotics that indirectly interfere with nucleic acid synthesis. Nucleic-acid synthesis can also be stopped by antibiotics that inhibit the enzymes that assemble these polymers, DNA polymerase or RNA polymerase.

Examples of such antibiotics are actinomycin, rifamicin, and rifampicin, the last two being particularly valuable in the treatment of tuberculosis.

The quinolone antibiotics inhibit synthesis of an enzyme responsible for the coiling and uncoiling of the chromosome, a process necessary for DNA replication and for transcription to messenger RNA. Some antibacterials affect the assembly of messenger RNA, thus causing its genetic message to be garbled. When these faulty messages are translated, the protein products are nonfunctional.

There are also other mechanisms: The tetracyclines compete with incoming transfer-RNA molecules; the aminoglycosides cause the genetic message to be misread and a defective protein to be produced; chloramphenicol prevents the linking of amino acids to the growing protein; and puromycin causes the protein chain to terminate prematurely, releasing an incomplete protein.

Range of Effectiveness

In some species of bacteria the cell wall consists primarily of a thick layer of peptidoglycan. Other species have a much thinner layer of peptidoglycan and an outer as well as an inner membrane. When bacteria are subjected to Gram's stain, these differences in structure affect the differential staining of the bacteria with a dye called gentian violet.

The differences in staining colouration (gram-positive bacteria appear purple and gram-negative bacteria appear colourless or reddish, depending on the process used) are the basis of the classification of bacteria into gram-positive (those with thick peptidoglycan) and gram-negative (those with thin peptidoglycan and an outer membrane), because the staining properties correlate with many other bacterial properties. Antibacterials can be further subdivided into narrow-spectrum and broad-spectrum agents.

The narrow-spectrum penicillins act against many gram-positive bacteria. Aminoglycosides, also narrow-spectrum, act against many gram-negative as well as some gram-positive bacteria. The tetracyclines and chloramphenicols are both broad-spectrum drugs because they are effective against both gram-positive and gram-negative bacteria.

Cell Death and Growth Suppression

Antibiotics may also be classed as bactericidal (killing bacteria) or bacteriostatic (stopping bacterial growth and multiplication). Bacteriostatic drugs are nonetheless effective because bacteria that are prevented from growing will die off after a time or be killed by the defence mechanisms of the host. The tetracyclines and the sulfonamides are among the bacteriostatic antiobiotics. Antibiotics that damage the cell membrane cause the cell's metabolites to leak out, thus killing the organism. Such compounds, including penicillins and cephalosporins, are therefore classed as bactericidal.

TYPES OF ANTIBIOTICS

Following is a list of some of the more common antibiotics and examples of some of their clinical uses. This section does not include all antibiotics nor all of their clinical applications.

Penicillins

Penicillins are bactericidal, inhibiting formation of the cell wall. There are four types of penicillins: the narrow-spectrum penicillin-G types, ampicillin and its relatives, the penicillinase-resistants, and the extended spectrum penicillins that are active against pseudomonas. Penicillin-G types are effective against gram-positive strains of streptococci, staphylococci, and some gram-negative bacteria such as meningococcus.

Penicillin-G is used to treat such diseases as syphilis, gonorrhea, meningitis, anthrax, and yaws. The related penicillin V has a similar range of action but is less effective. Ampicillin and amoxicillin have a range of effectiveness similar to that of penicillin-G, with a slightly broader spectrum, including some gram-negative bacteria. The penicillinase-resistants are penicillins that combat bacteria that have developed resistance to penicillin-G. The antipseudomonal penicillins are used against infections caused by gram-negative *Pseudomonas* bacteria, a particular problem in hospitals. They may be administered as a prophylactic in patients with compromised immune systems, who are at risk from gram-negative infections.

Side effects of the penicillins, while relatively rare, can include immediate and delayed allergic reactions—specifically, skin rashes, fever, and anaphylactic shock, which can be fatal.

Cephalosporin

Like the penicillins, cephalosporins have a Â-lactam ring structure that interferes with synthesis of the bacterial cell wall and so are bactericidal. Cephalosporins are more effective than penicillin against gram-negative bacilli and equally effective against gram-positive cocci. Cephalosporins may be used to treat strains of meningitis and as a prophylactic for orthopedic, abdominal, and pelvic surgery. Rare hypersensitive reactions from the cephalosporins include skin rash and, less frequently, anaphylactic shock.

Aminoglycosides

Streptomycin is the oldest of the aminoglycosides. The aminoglycosides inhibit bacterial protein synthesis in many gram-negative and some gram-positive organisms. They are sometimes used in combination with penicillin. The members of this group tend to be more toxic than other antibiotics. Rare adverse effects associated with prolonged use of aminoglycosides include damage to the vestibular region of the ear, hearing loss, and kidney damage.

Tetracyclines

Tetracyclines are bacteriostatic, inhibiting bacterial protein synthesis. They are broad-spectrum antibiotics effective against strains of streptococci, gram-negative bacilli, rickettsia (the bacteria that causes typhoid fever), and spirochetes (the bacteria that causes syphilis). They are also used to treat urinary-tract infections and bronchitis. Because of their wide range of effectiveness, tetracyclines can sometimes upset the balance of resident bacteria that are normally held in check by the body's immune system, leading to secondary infections in the gastrointestinal tract and vagina. Tetracycline use is now limited because of the increase of resistant bacterial strains.

Macrolides

The macrolides are bacteriostatic, binding with bacterial ribosomes to inhibit protein synthesis. Erythromycin, one of the macrolides, is effective against gram-positive cocci and is often used as a substitute for penicillin against streptococcal and pneumococcal infections. Other uses for macrolides include diphtheria and bacteremia. Side effects may include nausea, vomiting, and diarrhea; infrequently, there may be temporary auditory impairment.

Sulfonamides

The sulfonamides are synthetic bacteriostatic, broad-spectrum antibiotics, effective against most gram-positive and many gram-negative bacteria. However, because many gram-negative bacteria have developed resistance to the sulfonamides, these antibiotics are now used only in very specific situations, including treatment of urinary-tract infection, against meningococcal strains, and as a prophylactic for rheumatic fever. Side effects may include disruption of the gastrointestinal tract and hypersensitivity.

PRODUCTION

The production of a new antibiotic is lengthy and costly. First, the organism that makes the antibiotic must be identified and the antibiotic tested against a wide variety of bacterial species. Then the organism must be grown on a scale large enough to allow the purification and chemical analysis of the antibiotic and to demonstrate that it is unique.

This is a complex procedure because there are several thousand compounds with antibiotic activity that have already been discovered, and these compounds are repeatedly rediscovered. After the antibiotic has been shown to be useful in the treatment of infections in animals, larger-scale preparation can be undertaken.

Commercial development requires a high yield and an economic method of purification. Extensive research may be needed to increase the yield by selecting improved strains of the organism or by changing the growth medium. The organism is then grown in large steel vats, in submerged cultures with forced aeration. The naturally fermented product may be modified chemically to produce a semisynthetic antibiotic. After purification, the effect of the antibiotic on the normal function of host tissues and organs (its pharmacology), as well as its possible toxic actions (toxicology), must be tested on a large number of animals of several species.

In addition, the effective forms of administration must be determined. Antibiotics may be topical, applied to the surface of the skin, eye, or ear in the form of ointments or creams. They may be oral, or given by mouth, and either allowed to dissolve in the mouth or swallowed, in which case they are absorbed into the bloodstream through the intestines. Antibiotics may also be parenteral, or injected intramuscularly, intravenously, or subcutaneously; antibiotics are administered parenterally when fast absorption is required.

In the United States, once these steps have been completed, the manufacturer may file an Investigational New Drug Application with the Food and Drug Administration

(FDA). If approved, the antibiotic can be tested on volunteers for toxicity, tolerance, absorption, and excretion.

If subsequent tests on small numbers of patients are successful, the drug can be used on a larger group, usually in the hundreds. Finally a New Drug Application can be filed with the FDA, and, if this application is approved, the drug can be used generally in clinical medicine. These procedures, from the time the antibiotic is discovered in the laboratory until it undergoes clinical trial, usually extend over several years.

RISKS AND LIMITATIONS

The use of antibiotics is limited because bacteria have evolved defences against certain antibiotics. One of the main mechanisms of defence is inactivation of the antibiotic. This is the usual defence against penicillins and chloramphenicol, among others. Another form of defence involves a mutation that changes the bacterial enzyme affected by the drug in such a way that the antibiotic can no longer inhibit it. This is the main mechanism of resistance to the compounds that inhibit protein synthesis, such as the tetracyclines.

All these forms of resistance are transmitted genetically by the bacterium to its progeny. Genes that carry resistance can also be transmitted from one bacterium to another by means of plasmids, chromosomal fragments that contain only a few genes, including the resistance gene.

Some bacteria conjugate with others of the same species, forming temporary links during which the plasmids are passed from one to another. If two plasmids carrying resistance genes to different antibiotics are transferred to the same bacterium, their resistance genes can be assembled onto a single plasmid. The combined resistances can then be transmitted to another bacterium, where they may be combined with yet another type of resistance. In this way, plasmids are generated that carry resistance to several different classes of antibiotic. In addition, plasmids have evolved that can be transmitted from one species of bacteria to another, and these can transfer multiple antibiotic resistance between very dissimilar species of bacteria.

The problem of resistance has been exacerbated by the

use of antibiotics as prophylactics, intended to prevent infection before it occurs. Indiscriminate and inappropriate use of antibiotics for the treatment of the common cold and other common viral infections, against which they have no effect, removes antibiotic-sensitive bacteria and allows the development of antibiotic-resistant bacteria. Similarly, the use of antibiotics in poultry and livestock feed has promoted the spread of drug resistance and has led to the widespread contamination of meat and poultry by drug-resistant bacteria such as Salmonella.

In the 1970s, tuberculosis seemed to have been nearly eradicated in the developed countries, although it was still prevalent in developing countries. Now its incidence is increasing, partly due to resistance of the tubercle bacillus to antibiotics. Some bacteria, particularly strains of staphylococci, are resistant to so many classes of antibiotics that the infections they cause are almost untreatable. When such a strain invades a surgical ward in a hospital, it is sometimes necessary to close the ward altogether for a time. Similarly, plasmodia, the causative organisms of malaria, have developed resistance to antibiotics, while, at the same time, the mosquitoes that carry plasmodia have become resistant to the insecticides that were once used to control them. Consequently, although malaria had been almost entirely eliminated, it is now again rampant in Africa, the Middle East, Southeast Asia, and parts of Latin America. Furthermore, the discovery of new antibiotics is now much less common than in the past.

MOLD

Mold (fungi), fuzzy, cobweblike growth produced on organic matter by several types of fungi. Mold and mildew are commonly used interchangeably, although mold is often applied to black, blue, green, and red fungal growths, and mildew to whitish growths.

Black bread mold, *Aspergillus niger*, one of the most familiar molds, begins as a microscopic, airborne spore that germinates on contact with the moist surface of nonliving organic matter. It spreads rapidly, forming the mycelium (fungal body), which is made up of a fine network of filaments (hyphae). The mycelium produces other clusters of rootlike hyphae, called rhizoids, which penetrate the organic material, secreting enzymes and absorbing water and the digested sugars and starches. Other clusters of hyphae called sporangiophores then reach upward, forming sporangia (knoblike spore cases), which bear the particular colour of the mold species. Upon ripening, the sporangia break open and the windborne spores land elsewhere to reproduce asexually.

Some molds also reproduce sexually through conjugation of gamete cells by the joining of two specialized hyphae. The resulting zygote matures into a zygospore that germinates after a dormant period.

Molds thrive on a great many organic substances and, provided with sufficient moisture, they rapidly disintegrate wood, paper, and leather. In fruit the enzymes penetrate well behind the area of the visible growths to damage the fruit. Besides being destructive, however, molds also have many industrial uses, such as in the fermentation of organic acids and cheeses. Camembert and Roquefort cheeses gain their particular flavours from the enzymes of *Penicillium camemberti* and *P. roqueforti,* respectively. Penicillin, a product of the green mold *P. notatum,* revolutionized antibiotic drugs after its discovery in 1929, and the red bread mold *Neurospora* is an important tool in genetic experiments.

SLIME MOLDS

Slime Molds, multicellular organisms that form tiny, spore-producing fruiting bodies of various shapes. The two subgroups are the true slime molds, myxomycetes, and the cellular slime molds, acrasiomycetes.

True Slime Molds

Myxomycetes are most often found in moist, decaying leaf litter and logs in the forest. About 500 species can be distinguished by the size and shape of their fruiting bodies; some are delicately stalked, some branched, some show large masses of spores surrounded by a hard wall. Their life cycle is invariably sexual; two single cells of opposite sex fuse. The cells may be flagellated or amoeboid depending on whether the soil is moist or dry. After fertilization the nucleus divides, but no internal cell walls are formed, resulting in a multinucleate mass of protoplasm called a plasmodium. The plasmodium becomes a large, often orange, glistening mass of mucuslike material that slowly moves over the rotten wood or humus.

This mass is the feeding structure, which engulfs and ingests food (usually bacteria). When food becomes scarce and external temperature and humidity are right, the plasmodium transforms in a matter of hours into a series of bumps that become the fruiting bodies or sporangium. The spores are produced in the final stages of this differentiation. When the dispersed spores germinate, each one liberates a gamete.

Cellular Slime Molds

Acrasiomycetes are smaller than most myxomycetes and have fewer species (about 50). The cells are always amoeboid and uninucleate; they do not form plasmodia. Cellular slime molds constitute a large portion of the soil amoebae population, feeding separately on bacteria.

When they have cleared an area of food, they form fruiting bodies, but in a radically different manner from the myxomycetes. Starved amoebae stream together into central collection points, attracted by a chemical called acrasin that they themselves give off.

The resulting multicellular organism seems remarkably well organized: It has anterior and posterior ends, moves toward light, and orients toward or away from heat, depending on the temperature. When it comes to rest, the amoebae at the anterior end start producing a delicate stalk, while those at the posterior end turn into spores that form a terminal ball at the tip of the new stalk.

This cycle is asexual, although the cellular slime molds also have a sexual cycle. The fusion of the sex cells (individual amoebas) produces a large resistant body called a macrocyst, which upon germination, produces new cells capable of continuous asexual reproduction.

PROTISTA

Protista, group of comparatively simple organisms, called protists, that have characteristics of both plants and animals. Most protists are *unicellular* (consisting of a single cell) and can only be seen with a microscope, although there are some that are composed of more than one cell. There are a wide variety of protists, and they inhabit many different environments—fresh water, seawater, soils, and the intestinal tracts of animals, where they perform crucial digestive processes. Like plants, many species of protists can make their own food by the process of photosynthesis. Like animals, many protists can move around under their own power. Unlike plants and animals, however, protists do not have cells organized into specialized tissues.

The protists include such familiar organisms as seaweeds, amoebas, and slime molds. The kingdom Protista contains many economically important members, including organisms that cause diseases, such as malaria. Biologists theorize that members of the kingdom Protista gave rise to the kingdoms Plantae, Animalia, and Fungi about 600 million years ago.

All protists are *eukaryotes*. This means that their cells contain a *nucleus*, a membrane-bounded structure that encloses the cell's genetic material. (Organisms without nuclei—the bacteria and cyanobacteria, or blue-green algae—are called prokaryotes.) Although most protists have a single nucleus, protists are unique in that some contain multiple nuclei—up to ten thousand in a single cell—and others, such as ciliates, have two different-sized nuclei in a single cell. The deoxyribonucleic acid (DNA) of protists (like that of other eukaryotes) is organized in long molecules called chromosomes within the nucleus.

When the cell divides, these chromosomes *replicate* (duplicate themselves) and then divide in a process called *mitosis*. Among some protists, the *nuclear membrane* (the thin layer around the nucleus) stays intact during mitosis, whereas in plants, animals, and fungi the nuclear membrane breaks down.

Protists vary greatly in size and shape. Many are minute: The green alga *Nanochlorum* is only 0.01 mm (0.0004 in) long. Giant kelps can grow to 65 m (210 ft) or more in length. Some protists form structures known as spores that are resistant to chemicals and drying and that disperse in the environment. Spores are often the infectious stage of organisms that cause diseases in humans when ingested.

Anatomy and Physiology

Most protists contain many *mitochondria,* membrane-bounded organelles (cellular parts) that break down complex organic molecules and, in the process, release the chemical energy that powers the rest of the cell. The physiology of these protists is very similar to the cellular physiology of plants, animals, and fungi. Mitochondria require oxygen. Protists that live in environments without oxygen—for instance, in black muds or inside the digestive tract of animals—generally do not have mitochondria, and they create cellular energy by processes unique to the kingdom Protista.

In addition to mitochondria and nuclei, membrane-bounded organelles called *plastids* are found in protists that perform photosynthesis. Plastids capture the energy of sunlight and convert the energy into a chemical form available for use in the cell. Plastids are also able to capture inorganic carbon dioxide from the atmosphere and turn it into organic carbon useful for cell growth. Pigments, including chlorophyll, synthesized within plastids capture the sunlight and give photosynthetic protists their distinct colours: The brown algae contain brown plastids; the green algae, green plastids; and the red algae, red plastids.

In all likelihood, plastids and mitochondria were once free-living bacteria that became organelles through the process of symbiosis (an intimate and ongoing association of two or more different species). Once controversial, the theory of the symbiotic origin of plastids and mitochondria is generally accepted today. DNA molecules within present-day plastids and mitochondria are probably relics of the DNA of free-living bacteria ancestors.

Some protists are capable of movement. Protists known as flagellates move by means of *flagella,* long whiplike structures that the protists beat to propel themselves through water. Other protists use numerous and comparatively short structures called cilia. One group of protists, the ciliates, are distinguished by the great number and unusual arrangement of cilia along their surface. The single-celled amoebas and their multicellular relatives the slime molds are protists that use *pseudopodia,* temporary protrusions of the cell, both for movement and to engulf prey.

The vast majority of protists reproduce asexually—cells simply grow and divide. *Sexual recombination* (the production of gametes and the merging of DNA from two individuals to form a zygote) has been observed in some protists. However, many protists have been only poorly studied.

Classification

The first detailed descriptions of protists were made in 1676 by the inventor of the microscope, Dutch naturalist Antoni van Leeuwenhoek, who observed microscopic organisms that he called animalcules. Traditionally all organisms that moved had been considered animals, whereas all photosynthetic organisms had been considered plants. The term *Protista* was first used in 1862 by the German biologist Ernst Haeckel to describe microscopic organisms that were neither clearly animallike nor plantlike.

Haeckel recognized that some organisms are both motile and photosynthetic. For almost a hundred years after Haeckel, scientists informally recognized three kingdoms: plants, animals, and protists. The Protista included all the microbes. In the 1930s, it was formally proposed that all single-celled organisms, including bacteria, be placed in their own kingdom, Kingdom Protista.

In 1959 American biologist R. H. Whittaker described a classification system of five primary kingdoms: plants, animals, fungi, protists, and bacteria. Because the Protista are so diverse in form, classification within the kingdom has proved difficult. The classification of the Protista is currently based largely on the structure and organization of the cell, the presence of organelles, and the pattern of reproduction or life cycles. The five-kingdom classification system divides the Protista into 27 distinct phyla. More recently, however, classifications based on comparisons of cell physiology and DNA sequences suggest that many protist phyla may be sufficiently large and diverse to be classified as kingdoms.

The plantlike protists include the golden algae (phylum Chrysophyta), dinoflagellates (Pyrrophyta), cryptomonads (Cryptophyta), and euglenoids (Euglenophyta). The animallike protists, which are also called protozoa, include the animal flagellates (Zoomastigina), amoeboid forms (Sarcodina), ciliates and suctorians (Ciliophora), and the parasitic, spore-producing sporozoans (Sporozoa).

Funguslike forms include the hyphochytrids (Hyphochytridiomycota) and the plasmodiophores (Plasmodiophoromycota). The slime molds, which include several disputed phyla here treated as belonging to the Protista, have characteristics of both fungi and protozoans.

Chapter 6

The Importance of Fungi

Fungi are one of the most important groups of organisms on the planet. This is easy to overlook, given their largely hidden, unseen actions and growth. They are important in an enormous variety of ways.

RECYCLING

Fungi, together with bacteria, are responsible for most of the recycling which returns dead material to the soil in a form in which it can be reused. Without fungi, these recycling activities would be seriously reduced. We would effectively be lost under piles many metres thick, of dead plant and animal remains.

MYCORRHIZAE AND PLANT GROWTH

Fungi are vitally important for the good growth of most plants, including crops, through the development of mycorrhizal associations. As plants are at the base of most food chains, if their growth was limited, all animal life, including human, would be seriously reduced through starvation.

FOOD

Fungi are also important directly as food for humans. Many mushrooms are edible and different species are cultivated for sale worldwide. While this is a very small proportion of the actual food that we eat, fungi are also widely used in the production of many foods and drinks. These include cheeses, beer and wine, bread, some cakes, and some soya bean products.

While a great many wild fungi are edible, it can be difficult to correctly identify them. Some mushrooms are deadly if they are eaten. Fungi with names such as 'Destroying Angel' and 'Death Cap' give us some indication that it would not be a terribly good idea to eat them! In some countries, collecting wild mushrooms to eat is a popular activity. It is always wise to be totally sure that what you have collected is edible and not a poisonous look-a-like.

MEDICINES

Penicillin, perhaps the most famous of all antibiotic drugs, is derived from a common fungus called *Penicillium*. Many other fungi also produce antibiotic substances, which are now widely used to control diseases in human and animal populations. The discovery of antibiotics revolutionized health care worldwide.

Some fungi which parasitise caterpillars have also been traditionally used as medicines.

The Chinese have used a particular caterpillar fungus as a tonic for hundreds of years. Certain chemical compounds isolated from the fungus may prove to be useful treatments for certain types of cancer.

A fungus which parasitises Rye crops causes a disease known as Ergot. The fungus can occur on a variety of grasses. It produces small hard structures, known as sclerotia. These sclerotia can cause poisoning in humans and animals which have eaten infected material. However, these same sclerotia are also the source of a powerful and important drug which has uses in childbirth.

BIOCONTROL

Fungi such as the Chinese caterpillar fungus, which parasitise insects, can be extremely useful for controlling insect pests of crops. The spores of the fungi are sprayed on the crop pests. Fungi have been used to control Colourado potato beetles, which can devastate potato crops. Spittlebugs, leaf hoppers and citrus rust mites are some of the other insect pests which have been controlled using fungi. This method is generally cheaper and less damaging to the environment than using chemical pesticides.

CROP DISEASES

Fungal parasites may be useful in biocontrol, but they can also have enormous negative consequences for crop production. Some fungi are parasites of plants. Most of our common crop plants are susceptible to fungal attack of one kind or another. Spore production and dispersal is enormously efficient in fungi and plants of the same species crowded together in fields are ripe for attack. Fungal diseases can on occasion result in the loss of entire crops if they are not treated with antifungal agents.

ANIMAL DISEASE

Fungi can also parasitise domestic animals causing diseases, but this is not usually a major economic problem. A wide range of fungi also live on and in humans, but most coexist harmlessly. Athletes foot and Candida infections are examples of human fungal infections.

FOOD SPOILAGE

It has already been noted that fungi play a major role in recycling organic material. The fungi which make our bread and jam go mouldy are only recycling organic matter, even though in this case, we would prefer that it didn't happen! Fungal damage can be responsible for large losses of stored food, particularly food which contains any moisture. Dry grains can usually be stored successfully, but the minute they become damp, moulds are likely to render them inedible. This is obviously a problem where large quantities of food are being produced seasonally and then require storage until they are needed.

FUNGI AND THEIR HABITATS

Fungi can be found in just about any habitat you care to mention, from sea water through to freshwater, in soil, on plants and animals, on human skin and even growing on microscopic crevices in CD-ROM disks!

Most of the fungi you would find in these places however, are very small and would need a microscope or magnifying lenses to see. The following discussion on fungal habitats concentrates on macrofungi, that is, those with fruiting bodies big enough to be easily seen.

- Woods and meadows are the best places to go hunting for fungi. Of the two, woods are by far the best place to look, as over 80% of fungi are associated with trees.
- Some fungi, such as the Fairy Ring Toadstool, *Marasmius oreades,* and the edible field mushroom, *Agaricus campestris,* prefer open, grassy places.
- Also watch out for a different group of fungi which grow on animal dung. These tend to have very specific habitat requirements and you will usually find that particular species will grow only on a certain kind of dung, at a particular stage of decay. In order to identify the fungus, you may first need to identify the type of dung!
- Fungi are common in woodlands because of all the rotting wood and leaf litter, which provides a wide range of the dead organic matter which most fungi feed on.
- Many fungi are also associated with trees because they are linked into the tree roots. The association benefits both the fungi and the trees. This particular type of association between fungi and the roots of plants such as trees, is known as a mycorrhiza.
- Some fungi have very specific associations and will grow only with one kind of tree the bolete, *Uloporus lividus,* grows only under alders.
- Other fungi may be found in association with several different trees. Chanterelles, can be found linked with birch, pine, oak and beech trees. .
- Soil type is also important. Some fungi may be associated with a particular tree, but only where it is growing on suitable soil. The most important factor is usually whether the soil is acidic or calcareous

(chalky). Soil fertility also plays a part. Fields which have been heavily fertilised with artificial nitrates are less likely to be good mushroom hunting territory than those which have been organically fertilised.

MYCORRHIZAE

- Mycorrhizae are associations between fungal hyphae and the roots of plants. Almost all terrestrial plants, including wild plants, trees and commercial crops will form mycorrhizal associations with fungi in the soil. These mycorrhizae are vitally important for the growth and health of the plants.
- The fungi extract food (sugars) that they need from the plants, but in return they supply the plants with some of the nutrients and water which they may require. Thus both the fungi and the plants flourish because of the association. This is known as a symbiotic relationship. Certain plants, such as orchids, are totally dependent on a fungus associated with their roots in order to grow at all.
- Through their enormous collecting network of hyphae in the soil, the fungi help to supply nutrients, especially phosphorus, to the plant roots they are associated with. This can have amazing effects on plant growth, particularly in soils which are not very fertile. In soils containing little phosphorus, plants with mycorrhizae have been shown to grow up to 20 times faster than those without.
- The survival of plant seedlings may also be up to five times greater if they have mycorrhizae to help them collect nutrients and water from the soil. Helping plants to obtain phosphorus from the soil may also give them increased drought tolerance, as this is one of the effects of improved phosphorus nutrition.
- Mycorrhizal fungi can often form associations with many different kinds of plants at the same time. The fungal hyphae of several different kinds of fungi may

therefore be like a giant underground network connecting most of the plants in a habitat together. It is possible, although not proven, that this enables different plants to exchange nutrients between them via the fungal hyphae. If this is so then this would hugely improve the chances of seedlings surviving, because they are not just dependent on their tiny root system, but have access to the great underground collecting network.

- The presence of a good network of mycorrhizal fungi in the soil is therefore of vital importance for good plant growth in most habitats. Where the fungi are absent, or only present as isolated spores in the soil, plant growth will be reduced, apart from those species which do not require mycorrhizae. Most of these are what are commonly termed weed species.
- Mycorrhizae also help to develop good soil structure through production of a protein which helps to stick small particles of soil together to form larger ones. This means that water can move more easily through the soil and provides more air spaces and thus air, for soil organisms and plant roots.
- There are several different kinds of mycorrhizae. In some, the fungal hyphae actually enter the cells of the plant roots. These are called endomycorrhizae. There are several different kinds of these endomycorrhizae, of which a type abbreviated to AM (Arbuscular Mycorrhizae), is perhaps the best known. Eighty percent of the world's plant species, from grasses, to trees, most crop plants, shrubs and flowers, form AM associations.
- Another type is called an Ectomycorrhiza (ECM). In this kind, the fungus forms associations with plant roots, but does not actually enter the root cells. This kind is mostly formed with different types of trees, such as pines, firs, spruces and oaks, amongst many others. ECM fungi may form thick hyphal strands known as rhizomorphs which can conduct water

and nutrients relatively long distances. Many of the fungi common in woodland are ectomycorrhizal fungi, with mushroom or toadstool-like fruiting bodies.

FUNGI

Fungi are eukaryotic organisms distinct from plants and animals and members of several other smaller kingdoms. Common fungi include mushrooms, conks, corals, jellies, puffballs, stinkhorns, morels, cups, truffles, lichens, yeasts, rusts, smuts, bread molds, mildews, and molds on bathroom tiles.

In 1959, R. H. Whittaker introduced a five-kingdom taxonomy that granted fungi equal status with plants and animals. The five-kingdom system has been supplanted by a multiple-kingdom classification, and species traditionally treated as fungi are now distributed across several kingdoms. Those believed to form a monophyletic lineage are assigned to kingdom Eumycota (often called kingdom Fungi). Mycology, the science devoted to fungi, still covers all traditional fungi.

Characteristics of Fungi

The Eumycota consist of eukaryotic, nonchlorophyllous heterotrophs that absorb nutrients from dead or living organic matter, have cell walls composed of chitin, and store excess energy as glycogen. The kingdom contains four phyla: Chytridiomycota, Zygomycota, Ascomycota, and Basidiomycota. All true fungi have a definite cell wall throughout all developmental stages. Fungal cell walls are composed of chitin, the compound also found in arthropod exoskeletons (example, lobster shells). Most fungi produce a vegetative mycelium (filamentous thallus) composed of hyphae that branch and extend via tip elongation, although some groups (like yeasts) consist only of individual cells. Hyphae (singular, hypha) are tube-like filaments with either single multinucleate cells (coenocytes) that lack septa (cross-walls) separating nuclei, or many septate cells containing one, two, or more nuclei.

Fungal Nutrition: Saprobes, Parasites, and Mutualists

A fungal thallus may be as small as a single microscopic cell (baker's yeast, *Saccharomyces cerevesiae*) or exceedingly large (*Armillaria gallica*, the several-acre-sized "humungous fungus" reported in 1992 as the world's largest organism). Sporocarp (fruit body, or "mushroom" size) also ranges from microscopic to meters in diameter (*Bridgeoporus nobilissimus*, an endangered bracket fungus found on noble fir trees).

Not unexpectedly, such a diverse kingdom manifests several different life cycles. Virtually all fungi produce spores. Both asexual and sexual spores may germinate to form vegetative thalli from primary and secondary mycelia. Thalli may be haploid dominant, diploid dominant, or exhibit haplo-diploid alternation of generations. Here it is important not to confuse the chromosomal state of individual nuclei (haploid versus diploid) with the number of nuclei per cell (monokartyotic versus dikaryotic). Fungi are unusual in that they often exhibit dikaryotamy, wherein hyphal cells contain two (usually haploid) nuclei that migrate, multiply, and divide together.

Although superficially similar to plants, fungi are probably more closely related to animals. Like animals, fungi lack chlorophyll and do not photo-synthesize, must obtain nutrients from organic sources, and store energy as glycogen instead of starch. Unlike animals, however, fungi do not engulf, but rather absorb, their nutrients after breaking them down via enzymatic action, earning them the nickname "absorbotrophs."

Fungi absorb their nutrients in three different ways: (1) saprobes decompose dead organic matter; (2) parasites feed on living hosts; and (3) mutualists live in symbiotic unions with other living organisms. Saprophytic fungi, such as edible meadow mushrooms (*Agaricus campestris*), shiitake (*Lentinula edodes*), and oyster mushrooms (*Pleurotus ostreatus*), decompose dead plant and animal tissue by releasing enzymes from hyphal tips, thereby recycling organic materials back into the surrounding environment. Parasitic fungi also use enzymes to break down living tissue, usually sapping the energy of the host and frequently causing its demise.

Lichens and mycorrhizae are two important mutualistic associations. Lichens represent partnerships between a fungus (mycobiont) and one or more algae (phycobiont). Although there are a few basidiolichens, almost all lichen mycobionts are Ascomycota (approximately 20,000 ascolichens described thus far). Mycorrhizae are symbiotic or non to slightly pathogenic fungus-plant unions formed with approximately 85 percent of the vascular plants.

Mycorrhizae are identified as ectomycorrhizal, arbuscular mycorrhizal (AM), ericoid, orchid, arbutoid, and monotropoid based on anatomical form and association. Ectomycorrhizal fungi (predominantly basidiomycetes such as boletes, amanitas, and coral fungi) form thick mycelial mantles around rootlets of many trees (oaks, firs, pines, poplars) to which they transport water and minerals from the soil, receiving sugars and other organic nutrients in return. AM fungi (in the Zygomycota order Glomales) form an endo-infection by penetrating rootlets to form coils and vesicles or finely branched arbuscules. The last four mycorrhizal types are specific to individual plant groups.

ECOLOGICAL AND ECONOMIC IMPORTANCE

Fungi have a profound biological and economic impact. As decomposers, plant pathogens, and symbiotic partners, their ability to grow anywhere, on anything, makes them both beneficial and harmful recyclers of carbon and nitrogen. Beneficially, they are used as food (mushrooms, truffles) and in baking and brewing (yeasts). They are being developed to detoxify pollutants (soil fungi), control insects (pathogenic Zygomycota), and regulate plant growth (pathogenic Ascomycota). Detrimentally, rusts, smuts, and molds cost billions of dollars through crop disease and spoilage while forest pathogens such as the honey mushroom (*Armillaria ostoyae*) and root-butt rot (*Heterobasidion annosum*) similarly threaten the timber industry. Some are toxic when eaten, such as the infamous destroying angel (*Amanita phalloides*).

Natural LSD, a hallucinogen produced by ergot (*Claviceps purpurea*), is associated with medieval hysterical frenzies produced by consumption of infected grain, and the aflatoxin produced by *Aspergillus flavus* in improperly stored grain is one of the most potent carcinogens yet discovered. As human and animal pathogens, fungi cause infections that range from the vexing (athlete's foot, yeast infections) to life threatening (histoplasmosis). Fortunately, other fungi (such as *Penicillium*) have been used to develop modern antibiotics and beneficial immunosuppressants.

Classification

With the introduction of deoxyribonucleic acid (DNA) sequence analyses, previous fungal evolutionary theory is undergoing rapid transformation, not surprising when one considers the extremely fragmentary fungal fossil record. Mycologists generally agree that true fungi (with animals and plants) diverged from the protozoan fungi (example, slime molds and Oomycota) before fungi and animals diverged from plants.

The chytrids separated from the remaining phyla approximately 550 million years ago, followed by splintering of the Zygomycota from the other two phyla, which share septate hyphae and dikaryotic stages in their life cycles.

The Ascomycota split from the Basidiomycota approximately 400 million years ago, followed by an increase in fungal diversity throughout the Paleozoic. Most yeasts and "asco-molds" are thought to have evolved in approximately the last 200 million years.

While most plants and vertebrates appear to have been described, most fungi have yet to be discovered or named. The 1995 edition of the *Dictionary of Fungi* counts 566,360 described species (about 100,000 in the Eumycota) but notes that there may be approximately 1.5 million species of fungi in the world. Elias Fries (1794–1878), often regarded as the "father of mycology," classified fungi based primarily on spore print colour and sporocarp appearance. Friesian-based nomenclature, reflecting similarity of form (phenetic) rather than genetic relationships, is still used in many field guides. Now, however, taxonomists are integrating molecular and morphological characters to develop natural classifications that more adequately reflect evolutionary relationships. Realizing that fungal taxonomy and nomenclature will remain somewhat fluid until new species and data are analysed and integrated, most mycologists generally accept the classification below.

Chytridiomycota.

The fact that chytrids alone among the Eumycota produce motile zoospores explains why their phylum is sometimes assigned with the flagellate oomycetes to kingdom Chromista. Chytrids possess posteriorly uniflagellate spores, mitochondria with flattened cristae, and cell walls composed of glucan and chitin. Among the simplest and smallest fungi, they live as saprobes in water and damp organic-rich habitats, or as parasites on invertebrates, plants, and other fungi. The so-called "frog chytrids" (*such as Batrachochytrium dendrobatidis*) are implicated in the current worldwide decline of amphibian species. Other chytrids are host-specific "rumen fungi" (*such as Piromyces communis*) that thrive in anaerobic conditions in the guts of herbivores, such as cattle and sheep. There are five orders and about 800 species of chytrids recognized thus far.

Unicellular members of the order Chytridiales lack a mycelial stage and consist of a central body with a few rudimentary appendages (haustoria) that attach to and invade the host tissue. Other chytrids, such as the Blastocladiales, develop true mycelia with sporangia and male/female gametangia that produce uniflagellate zoospores.

Chytrid thalli may be either haploid or diploid, and some, like the aquatic chytrid *Allomyces,* exhibit an isomorphic alternation of generations with similar appearing sexual and asexual zoospores.

Zygomycota.

The chytrids and this phylum are assigned the two "bottom" branches of the fungal evolutionary tree. There are more than 1,000 species in two classes (Trichomycetes, Zygomycetes) and ten orders, representing a diverse assemblage of saprobes, soil fungi, obligate insect and fungal parasites, and mycorrhizal formers. Common representatives of the saprobic order Mucorales include the dung-inhabiting "cap-thrower" (*Pilobolus*), the black bread mold (*Rhizopus stolonifera*), and *Phycomyces blakesleeanus*, sometimes referred to as the "body in the basement" because of the rapid growth of long, hairlike sporangiophores over a substrate under the right conditions.

Zygomycota are characterized by large, thick-walled, coenocytic zygospores and hyphae with relatively thin walls composed of chitin and chitosan. Both asexual sporangiospores and sexual zygospores germinate into haploid mycelia, with the hyphae functioning as gametangia during the sexual stage. In *Rhizopus*, for instance, close proximity of two hyphal strands of different mating types chemically triggers each to grow branches toward the other to form septate suspensor cells and gametangia. Eventual fusion produces a diploid zygosporangium that undergoes meiosis to become a thick-walled zygospore with large numbers of haploid nuclei.

Ascomycota.

In addition to most lichens and so-called "imperfect fungi," about 33,000 species of unicellular yeasts, green and black molds, powdery mildews, morels, cup fungi, and ascotruffles ("true" truffles) belong to this phylum. The phylum is characterized by ascospores produced within a saclike sporangium called an ascus. Mycelia (more complex than Zygomycota mycelia) are composed of septate hyphae with chitin-glucan hyphal walls. Most species produce specialized fruiting bodies called ascocarps whose details of structure help define different species, classes, or orders. Nonascocarpic representatives (such as unicellular yeasts and mildews that reproduce primarily by budding) do not form mycelia. Both sexual and asexual reproduction are found within this phylum.

Basidiomycota.

This phylum, which also features septate hyphae and chitin-glucan cell walls, is characterized by basidiospores borne upon a club-like structure called a basidium. Approximately 22,500 species are assigned to three classes:

Basidiomycetes, Teliomycetes, and Ustomycetes. Basidiomycetes include mushrooms, polypores, crusts, corals, clubs, basidiolichens, and jellies, which propel their spores, and "gastromycetes" (or "stomach fungi") that passively release their spores (puffballs, basidiotruffles ["false truffles"], stinkhorns, and birds' nests). Teliomycetes (rusts) and Ustomycetes (smuts) are obligate parasites of insects or plants. Rusts and smuts have exceedingly complex cycles involving up to five separate spore stages and multiple hosts. This ability to produce spores on different hosts in multiple ways presents a significant economic challenge to agriculture.

Oomycota.

Oomycetes (kingdom Chromista) are distinguished from true fungi by having glucan-cellulose cell walls that only occasionally incorporate small amounts of chitin. These algae-like fungi occur in aquatic or moist terrestrial habitats as single cells or mycelial mats composed of multinucleate, nonseptate hyphae. Their life cycle generally mirrors that of plants, with a transitory haploid stage. Both resting oospores and motile zoospores are diploid, the latter propelled by two unequal flagella (tinsel type plus whiplash). This phylum contains about 700 species in nine orders, including. The gills of a mushroom. Although superficially similar to plants, fungi are members of a distinct kingdom.

the generally saprophytic water molds (Saprolegniales) and the pathogenic Peronosporales and Pythiales. Most water molds are saprophytic, but there are a number of parasites that invade plants (white rusts, downy mildews, tobacco blue mold) or fish. Among the economically significant wilts, blights, and pathogens are *Phytophthora infestans* (responsible for the Irish potato famine in the 1850s), *Plasmopara viticola* (the causative agent of downy mildew of grapes), and the fish parasite *Saprolegnia parasitica,* a twenty-first-century threat to salmon migrating through dams.

ALTERNATION OF GENERATIONS

For sexually reproducing multicellular organisms such

as plants and animals, the life cycle requires that diploid cells divide by meiosis to create haploid cells. Haploid cells then fuse to recreate the diploid number and a new organism. Alternation of generations refers to the occurrence in the plant life cycle of both a multicellular diploid organism and a multicellular haploid organism, each giving rise to the other.

This is in contrast to animals, in which the only multicellular phase is the diploid organism (such as the human man or woman), whereas the haploid phase is a single egg or sperm cell.

The fern is an example of alternation of generations, in which both a multicellular diploid organism and a multicellular haploid organism occur and give rise to the other.

Alternation of generations is easiest to understand by considering the fern. The large, leafy fern is the diploid organism. On the undersurface of its fronds or leaves, its cells undergo meiosis to create haploid cells. However, these cells do not immediately unite with others to recreate the diploid state. Instead, they are shed as spores and germinate into small haploid organisms.

Because the diploid organism creates spores, it is called the sporophyte generation of the life cycle. Upon reaching maturity, the haploid organism creates haploid egg and sperm cells (gametes) by mitosis. Because the haploid organism creates gametes, it is called the gametophyte generation of the life cycle. The male gametes (sperm) are then released and swim to the female egg. Fusion of the gametes creates the new diploid sporophyte, completing the life cycle.

Whereas the fern gametophyte and sporophyte generations are completely independent, in some types of plants one generation lives on or in the other and depends on it for nutrition. In mosses, the familiar lush carpet of moss is the gametophyte, and its gametes require a moist environment for short-distance swimming before fusing. The sporophyte lives as a thin stalk attached to the gametophyte. Spores are released into the air and can travel on the slightest breezes to other habitats.

In contrast, in flowering plants (angiosperms), the

sporophyte is the dominant form. The male gametophyte has been reduced to just three cells, two of which are sperm. These together form the pollen grain, which is formed from the anther, part of the sporophyte. Similarly, the female gametophyte has been reduced to just seven cells, one of which is the egg cell. These are retained inside the ovule, which is part of the sporophyte. In angiosperms, two fertilization events take place: one sperm fertilizes the egg to form the diploid zygote of the new individual, and the other sperm fertilizes the so-called polar nuclei to form the triploid endosperm, a nutritive tissue. Together with maternal sporophyte tissue, these make up the seed.

MYCORRHIZAE

Symbioses are intimate associations between two unrelated organisms. Mycorrhizae are very common but largely unseen symbioses between plant roots and fungi that are important in plant nutrition, community structure, and nutrient cycling. Throughout the course of their evolution, plants and fungi have formed many different types of mycorrhizal partnerships involving most plant families and thousands of fungal species.

These diverse symbioses have been grouped into general types: arbuscular mycorrhizae, ectomycorrhizae, orchid mycorrhizae, and mycorrhizae. Endotrophic mycorrhizae in an orchid root. Mycorrhizae are very common but largely unseen symbioses between plant roots and fungi.

In plants in the order Ericales including ericoid, arbutoid, and monotropoid mycorrhizae.

Mycorrhizae are critical for the mineral nutrition of many plants because threadlike fungal hyphae can exploit soil much more extensively than plant roots, and thus mycorrhizal associations greatly increase the absorption of minerals and water. Usually, mycorrhizal fungi supply minerals to their host plants, which reciprocate by supplying carbohydrates to their fungal associates, but there are a few exceptions. In orchids, and some chlorophyll-free plants in the order Ericales, the flow of carbon is reversed, and mycorrhizal fungi supply the plant with organic carbon derived from dead plant matter or from neighbouring living plants.

Arbuscular mycorrhizae, the most common type, are associations between most crop plants, grasses, forbs, and many trees and fungi in the division Zygomycota, order Glomales. Both fossil and molecular evidence indicate that the earliest land plants had arbuscular mycorrhizal partnerships 450 million years ago. Ectomycorrhizae are commonly formed by woody shrubs and trees and a diverse array of fungi in the divisions Basidiomycota and Ascomycota. Pines and other forest trees often grow poorly or cannot survive in the absence of ectomycorrhizae.

Taxa of mycorrhizal fungi differ greatly in their effects on plant fitness. Consequently, interactions between communities of mycorrhizal fungi and plants may have strong impacts on the structure and function of communities and ecosystems.

SYMBIOSIS

A wide array of interactions among plants, animals, and microorganisms occurs in nature. Some of these relationships are characterized by a close physical association among species that persists for a significant period of the life cycle. In 1879 German botanist Heinrich Anton de Bary coined the term "symbiosis" to describe these relationships, meaning the living together of different species of organisms.

An interaction is considered a symbiosis based on the closeness of the physical association among the organisms rather than on the effect or outcome of the interaction. Symbiotic relationships span a spectrum from beneficial to detrimental effects. Many people associate symbiosis with mutualism, interactions that are beneficial to the growth, survival, and/or reproduction of *both* interacting species. But symbiotic interactions also include commensalism (one species receives benefit from the association and

Cleaner shrimp cleaning a zebra moray eel. Mutualistic relationships such as these promote the well-being of the host fishes and provide food for those that do the cleaning.

The other is unaffected), amensalism (one species is harmed, with no effect on the other), and parasitism. An example of commensalism is found in the anemone fish, which gains protection from living among the poisonous tentacles of the sea anemone, but offers no known benefit to its host.

In parasitic interactions, one species lives on or within a host organism and receives nourishment from the host, whereas the host is harmed by the interaction.

In obligate interactions, the relationship is essential to at least one of the interacting species. Facultative interactions are those that are beneficial to at least one of the interacting species, but not essential.

MUTUALISMS IN PLANTS

A common and widespread symbiosis occurs between terrestrial plants and fungi that colonize their roots. These associations are called "mycorrhizae," a word meaning "fungus-root." Unlike pathogenic fungi that cause disease, mycorrhizal fungi benefit the plant in several ways. These fungi germinate from spores in the soil to form thin threadlike structures called hyphae, which grow into the roots of plants. Once the roots are colonized, the fungal hyphae grow out from the root in an extensive network to explore the soil beyond the reach of the roots, gathering essential mineral nutrients and transporting them into the plant, increasing its growth. In return, the plant provides carbohydrates as a food source for the fungus.

Mycorrhizal symbiosis occurs in about 80 percent of all plant species. It is essential to many plants in low-nutrient environments because their roots alone are incapable of absorbing adequate amounts of some essential minerals such as phosphorus. The symbiosis is essential to the fungus because, unlike plants, fungi cannot make their own food via photosynthesis.

Mycorrhizal fungi provide other benefits to plants including improved resistance to drought and disease. The additional mineral nutrients acquired by these fungi have been shown to aid plants in coping with competitors and herbivores.

This symbiosis plays a large role in the growth and functioning of plants in both natural and agricultural ecosystems.

Legumes and certain other plants are colonized by *Rhizobium* bacteria that form small swellings or nodules on their roots. These symbiotic bacteria carry out the process of nitrogen fixation, the conversion of nitrogen gas into ammonia. Nitrogen is an essential element required by all organisms. Although nitrogen gas is abundant in the air, plants are unable to use nitrogen in this form, but they can readily use the ammonia formed by these bacteria and thus benefit from this symbiosis. As with mycorrhizal associations, the host plant benefits its symbiont by providing a carbohydrate energy source.

MUTUALISMS IN ANIMALS

In animals, a common mutualistic symbiosis occurs between many herbivores and microorganisms of their digestive tracts. Ungulates (hoofed animals) and some other animals eat plant material that is high in cellulose, even though they lack enzymes capable of breaking down cellulose molecules. They obtain energy from cellulose with the help of symbiotic bacteria and protozoa living within their digestive tracts. These microbes produce enzymes called cellulase that break down cellulose into smaller molecules that the host animal can then utilize. Similarly, wood-consuming termites depend upon symbiotic protozoans living within their intestines to digest cellulose.

These are obligate symbioses. The termites cannot survive without their intestinal inhabitants, and the microorganisms cannot live without the host. In each of these symbioses, the host animal benefits from the food provided by the microorganism and the microorganism benefits from the suitable environment and nourishment provided by the host.

A variety of animals engage in a mutualistic relationship referred to as cleaning symbioses. Birds such as oxpeckers benefit their large ungulate hosts by removing their external parasites, benefiting in return from the food source the host provides. In the marine environment, certain species of fish and shrimp similarly specialize in cleaning parasites from the outside of fishes. This mutualistic relationship promotes the well-being of the host fishes and provides food for those that do the cleaning. Unlike herbivores and their gut microorganisms, these interactions do not involve a close association of one organism living exclusively within another. These and other mutualistic but not clearly symbiotic relationships, such as those between plants and their pollinators, are sometimes referred to as proto-cooperation.

PARASITISM

Perhaps the most common type of symbiotic interaction in nature is parasitism. Many kinds of worms, protozoa, bacteria, and viruses are important animal parasites. Some, such as fleas or ticks, are ectoparasites, living on the outside of their host. Others, such as tapeworms or hookworms, are endoparasites that live inside their host.

A variety of parasitic symbionts also occur in plants. In some plants, insects deposit their eggs within the growing shoot tips or other plant part, at the same time producing chemicals that cause the development of a large swelling or tumorlike growth called a gall. The insect larvae then develop within the gall, feeding on the plant tissue as they grow. When its development is completed, the adult insect emerges from the gall to mate and then initiate the gall-forming cycle again. This is an obligate symbiosis because the insect larvae lives inside the plant and cannot complete its life cycle without its host plant. It is also a parasitic association because the insect living within the plant consumes plant tissue and causes harm to its host plant, while benefiting from the food resources and shelter provided by the plant. In addition to insects, other gall-forming symbionts include viruses, bacteria, and fungi.

Symbioses are widespread and important in the life of many organisms and ecologically important in the functioning of natural ecosystems. The patterns of adaptations of mutualists, parasites, and hosts suggest that these interactions are the product of coevolution, leading to increasingly specialized, and often increasingly beneficial, associations. In many mutualistic symbioses such as lichens (symbioses of algae and fungi) and corals (cnidarians and endosymbiotic algae), the adaptive value of the association is that one organism acquires from its partner some new metabolic capability (example, photosynthesis) that it does not itself possess.The condition of an organism which obtains its nourishment wholly or partially from the body of another living organism, and which usually brings about extensive modifications in both guest and host, a phenomenon widespread amongst animals and plants.

The term has been appropriated by biologists as a metaphor from the Greek. The lives of organisms are so closely intermeshed that if dependence on other organisms for food be the criterion of parasitism it is doubtful if any escape the taint.

Green plants, it is true, build up their food from the inorganic elements of the air and the soil, and are farthest removed from the suspicion of dependence; but most, if not all, thrive only by the aid of living microbes either actually attached to their roots or swarming in the nutrient soil. Saprophytes, organisms that live on organic matter, are merely parasites of the dead, whilst all animals derive their nourishment from the bodies of plants, either directly or indirectly through one or more sets of other animals. It is plain, therefore, that if parasitism is to be employed as a scientific term it must connote something more than mere dependence on another living organism for nutrition.

The necessary additional conceptions are two: the bodies of host and parasite must be in temporary or permanent physical contact other than the mere preying of the latter on the former; and the presence of the parasite must not be beneficial, and is usually detrimental to the host. It is obvious that within the limits of the strictest definition of parasitism that will cover the facts many degrees occur.

The terms *symbiosis* and *commensalism* have been applied to conditions really outside the definition of parasitism, but closely related and usually described in the same connexion. Both terms cover the physical consorting of organisms in such a fashion that mutual service is rendered.

The name symbiosis was invented by the botanist A. de Bary in 1879, and is applied to such an extraordinary community as the thallus of a lichen, which is composed of a fungus and an alga so intimately associated, physically and physiologically, that it was not until 1868 that the dual nature of the whole was discovered.

The presence of chlorophyll, which had always been associated only with vegetable organisms, was detected by Max Schultze in 1851 in the animals *Hydra* and *Vortex*, and later on by Ray Lankester in *Spongilla* and by P. Geddes in some Turbellarian worms. On the theory that the chlorophyll occurs in independent vegetable cells embedded in the animal tissues, such cases form other instances of symbiosis, for the oxygen liberated by the green cells enables their animal hosts to live in fouler water, whilst the hosts provide shelter and possibly nitrogenous food to their guests.

The term commensalism was introduced in 1876 by P. J. Van Beneden to cover a large number of cases in which "animals have established themselves on each other, and live together on a good understanding and without injury." The most familiar instance is that of fishes of the genus *Fierasfer* which live in the digestive tube of sea-cucumbers. A variety of commensalism was termed mutualism by Van Beneden and applied to cases where there appeared to be an exchange of benefits. A well-known instance of mutualism is the relation between sea-anemones and hermit crabs.

The hermit crab occupies the discarded shell of a mollusc, and anemones such as *Sagartia* or *Adamsia* are attached to the outside of the shell. The bright colours of the anemone advertise its distasteful capacity for stinging, and secure protection for the crab, whilst the anemone gains by vicarious locomotion and possibly has the benefit of floating fragments from the food of the crab.

It is plain that such terms as symbiosis, commensalism and mutualism cannot be sharply marked off from each other or from true parasitism, and must be taken as descriptive terms rather than as definite categories into which each particular association between organisms can be fitted.

R. Leuckart has made the most useful attempt to classify true parasites. Occasional, or temporary, parasites are to be distinguished from permanent, or stationary, parasites. The former seek their host chiefly to obtain food or shelter and are comparatively little modified by their habits when compared with their nearest unparasitic relatives.

They may infest either animals or plants, and as they attack only the superficial surfaces of their hosts, or cavities easy of access from the exterior, they correspond closely with another useful term introduced by Leuckart. They are Epizoa or Ectoparasites, as distinguished from Entozoa or Endoparasites. They include such organisms as plant-lice, and caterpillars which feed on the green parts of plants, and animals such as the flea, the bed-bug and the leech, which usually abandon their hosts when they have obtained their object. Many ectoparasites, however, pass their whole lives attached to their hosts; lice, for instance, lay their eggs on the hairs or feathers or in rugosities of the skin of birds and mammals; the development of the egg, the larval stages and the adult life are all parasitic.

Permanent or stationary parasites are in the most cases endoparasitic, inhabiting the internal organs; bacteria, gregarines, nematodes and tapeworms are familiar instances. But here also there are no sharp lines of demarcation. Leuckart divided endoparasites according to the nature and duration of their strictly parasitic life: (1) Some have free-living and self-supporting embryos that do not become sexually mature until they have reached their host; (2) others have embryos which are parasitic but migratory, moving either to another part of their host, to another host, or to a free life before becoming mature; (3) others again are parasitic in every stage of their lives, remaining in the same host, and being without a migratory stage.

ORIGIN OF PARASITISM

Now that the theory of spontaneous generation has been disproved, the problem of parasitism is no more than detection of the various causes which may have led organisms to change their environment. Every kind of parasite has relations more or less closely akin which have not acquired the parasitic habit, and every gradation exists between temporary and permanent parasites, between creatures that have been only slightly modified and those that have been profoundly modified in relation to this habit.

There are many opportunities for an animal or plant in its adult or embryonic stage to be swallowed accidentally by an animal, or to gain entrance to the tissues of a plant, whilst in the case of ectoparasites there is no fundamental difference between an organism selecting a dead or a living environment for food or shelter.

If the living environment in the latter case prove to have special advantages, or if the interior of the body first reached accidentally in the former case prove not too different from the normal environment and provide as better shelter, a more convenient temperature, or an easier food supply, the accident may pass into a habit. From the extent to which parasitism exists amongst animals and plants it is clear that it must have arisen independently in an enormous number of cases,. and it may be supposed that there must be many cases in which it has been of recent occurrence; E.

Metchnikoff, indeed, has suggested that amongst parasites we are to look for the latest products of evolution. In any case it is impossible to suppose that parasites form a natural group; no doubt in many cases the whole of a group, as for instance the group of tapeworms, is parasitic, but indications point clearly to the tapeworms having had free-living ancestors. Parasitism is in short a physiological habit, which theoretically may be assumed by any organism, and which actually has been assumed by members of nearly every living group.

Animals Vertebrat: These are rarely parasitic, and cases are unknown amongst mammals, birds, reptiles and amphibia. Amongst fish and cyclostomes, *Myxine* burrows into codfish, *Remora* attaches itself to the external surface of sharks; *Rhodens amarus,* the bitterling, a small, carp-like fresh-water fish, injects its eggs into the mantle-cavity of pond-mussels, where the fry develop, whilst the mollusc reciprocates by throwing off its embryos on the parent fish; *Stegophilus insidiosus,* a small colourless fish from Brazil and the Argentine, lives parasitically in the gill-cavity of large cat-fishes and sucks the blood in the gills of a large Silurid; *Vandellia cirrhosa,* the candiru of Brazil, a minute fish 60 mm. in length, enters and ascends the urethra of people bathing, being attracted by the urine; it cannot be withdrawn, owing to the erectile spines on its gill-covers.

The natives in some parts of the Amazon protect themselves whilst in the water by wearing a sheath of minutely perforated coco-nut shell.

MOLLUSCA

Few if any are true parasites. The Gasteropods, *Eulimae, Styliferae* and *Entoconchae* lodge in Echinoderms, the latter at least being truly parasitic.

Protochorda and *Hemichorda. - Most* of these are sessile and may lodge on other animals, but are not parasitic.

ARACHNIDA

Mites and *Ticks* are Arachnids, the vast majority of which are parasitic, and species of which infest almost every vertebrate group, but there are some free-living forms. Pycnogonids are parasitic in their youthful stages on Hydroids, whilst the Pentastomids have been so much modified by parasitism that they were long regarded as worms; they may occur in most vertebrates. *Crustacea. - These* contain an immense number of forms in all stages of parasitism. Some Copepods are amongst the most degenerate parasites known, the so-called fish-lice being for the most part Copepods with piercing mouth-organs, elaborate clinging apparatus, and degenerate organs of locomotion.

In *Lernea,* the female, after becoming attached to its host, undergoes a retrogressive metamorphosis, losing almost completely the segmentation of the body and discarding the appendages and sense-organs; whilst the male, although not so degenerate in structure, is dwarfed in size and itself becomes a parasite of the female.

The *Cirripeds* are all sessile in the adult condition. The

Lepadidae are the least modified and are rarely parasitic; the *Balanidae* are more modified and frequently become embedded in the skin of whales. The *Abdominalia* live as parasites buried in the shells of other Cirripeds and of molluscs.

The *Apoda* live as parasites in the mantle of other Cirripeds, whilst the *Rhizocephala* live chiefly on the abdomen of Decapod Crustacea, sending burrowing root-like nutritive processes into their tissues.

Insecta

A very large number of insects are temporary or permanent parasites of animals or plants, the adult stages being chiefly ectoparasitic, the larval stages endoparasitic. The *Hemimeridae,* allies of the earwigs, are ectoparasites on rats. The *Mallophaga* or bird-lice are degenerate wingless insects spending their whole lives as ectoparasites on birds and mammals. The larvae of *Hemerobiinae* are parasitic on Aphides. The saw-flies are parasitic on plants.

There are over 200,000 species known of the *Hymenoptera parasitica* or *Terebrantia.* The adults deposit their eggs in the eggs, caterpillars or adults of other insects, particularly Lepidoptera. The clothes-moth, for instance, is known to be subject to the attack of over sixty species of Hymenoptera. To such an extent has parasitism been developed in this group, that the parasites themselves are attacked by other parasites, giving rise to the phenomena known as hyperparasitism. The gall-flie are included amongst the Terebrantia, but in their case the early stages are passed in vegetable galls more frequently than in the bodies of other insects.

The ruby-flies (*Hymenoptera Tubulifera*), in the larval condition are parasitic on the larvae of wasps and bees. The *Denudatae* are bees that in the larval stage are parasitic on other bees, the larvae of the parasites being deposited in the food-cells prepared for their own larvae by other bees. Many of the fossorial Hymenoptera form no special nests for their young, but take advantage of the abodes and food-stores prepared by other insects.

The very large number of Hymenopterous insects that

collect living larvae to be shut up as provender for their developing young are in a sense parasitic. The complex relations of ants with other insects must be referred to in this connexion. The nests of many species are inhabited by foreign insects of various orders, such insects being termed myrmecophilous or ants'-nest insects.

The relations between the ants and their guests are very complex, and the guests migrate with their hosts. *Aphidae, Coccidae* and other bugs that secrete sugary matter are cherished and tended by ants; so also the caterpillars of some Lycaenid butterflies are kept as a kind of domesticated animal for some useful purpose.

There are also many Orthoptera, Hemiptera, and other insects, as well as some acarids and wood-lice found only in ants'-nests as cherished or tolerated guests.

The relations between ants and plants is also interesting; the ants live as parasites on the plants or trees, but in return protect them from more harmful intruders. Such phenomena are on the border-line between symbiosis and true parasitism. Although most beetles live on decaying animal or vegetable matter, a large number are parasitic in the adult or larval condition on animals or plants.

The curious beetle known as *Platypsyllus castoris* is known only as an ectoparasite of the beaver, whilst the *Leptinidae* are parasites of several species of mammals. The minute beetles of the families *Mordellidae* and *Rhipiphoridae* are endoparasites of wasps and cockroaches, whilst the larvae of many of the *Cantharidae* are parasites of locusts.

The *Strepsiptera* are endoparasites of Hymenoptera and Hemiptera. The habits of the Diptera easily pass over into parasitism, and a very large number are temporary or permanent parasites in the adult or larval stages. Most of the larvae of the *Cecidomyiidae* live in plants and form galls or other deformities. The bloodsucking habits of mosquitoes and gnats and sand-flies have not led to any special development in the direction of parasitism.

The larvae of *Bombyliidae* are endoparasites of the larvae of masonbees, and some of the *Cyrtidae* similarly infest spiders, whilst the *Tachinidae* deposit their larvae in other living insects, caterpillars being especially selected. The larvae of some of the *Sarcophagidae* may be deposited in the nostrils of man and other animals, where they may cause death, whilst those of the South American genus *Lucilia* infest the nasal fossae and frontal sinuses of man, producing great suffering, and the larvae of the numerous kinds of bot-fly attack man and many animals. The very large group of Pupipara live by sucking the blood of mammals and birds, and many of them are reduced to wingless permanent parasites.

The single member of the family *Braulidae* is a parasite of the bee. All the known fleas (Aphaniptera or Siphonaptera) are ectoparasites in the adult condition; the larval stages are usually to be found in organic refuse. The larvae of most Lepidoptera are temporary ectoparasites of plants, but a few attack other insects, such as coccids and aphids. All the Hemiptera (bugs) have sucking-mouth organs and the majority of them are temporary parasites of plants or other animals. Some, such as the bed-bug, have been so modified by parasitism as to be found only in human dwellings, others, such as the aphids or plant-lice, are permanent parasites of plants, many of them producing galls.

The coccids, or scale insects, have been still further modified as plant ectoparasites. The *Pediculidae,* or lice, are the most completely parasitic of insects, and are degraded wingless insects found on almost any kind of bird or mammal, but in most cases so highly modified as to be capable of existence only on the particular species. with which they are associated. *Lower Invertebrates. - No* true Chaetopods are parasitic, but a few are commensal.

The leeches are probably Chaetopods modified by parasitism; and Myzostomes are still more highly modified relatives of the group, very degenerate and parasitic on Crinoids. A few rotifers are ectoand endo-parasites. No Brachiopods, Polyzoa or Echinoderms are true parasites. The flat-worms and round-worms contain the most characteristic endoparasites, and parasitism is so characteristic a feature of most of the groups in the separate articles dealing with the various natural assemblages of such worms. All the Cestodes, most of the Trematodes, and a few of the Planarians are parasites of animals. Most Nemertines are free-living, but *Cephalothrix galatheae* is endoparasitic in the ovaries of the Crustacean *Galathea strigosa,* whilst *Eunemertes* and *Tetrastemma* occur on Ascidians, and *Malacobdella* in lamellibranch Molluscs.

The degraded Mesozoa are endoparasites of Planarians, Nemertines and Ophiurids. The Nematoda or typical round-worms, exhibit every degree from absolute free-life to absolute parasitism in animals and plants. The Echiuroidea are mostly free-living, but the male of *Bonellia* lives as a very degenerate parasite in the uterus and pharynx of the female. Although Coelentera and Porifera are usually sessile, very few are true parasites; young stages of the *Narcomedusae* are parasitic in the mouth of adults of different species, whilst *Mnestra parasites* is a degenerate medusa living on the pelagic mollusc *Phyllirhoe.*

The Protozoa, from their minute size and capacity to live in fluids, naturally include an enormous number of parasitic forms, the importance of which in producing disease in their hosts is so great that a very large special literature on parasitic protozoology is being formed. Of the *Sarcodina (q.v.)* many forms of Amoeba such as *Amoeba coli* are associated with dysentery and kindred diseases.

A very large number of the *Mastigophora,* including such forms as the trypanosome of sleeping-sickness, are parasitic; in fact, observation by adequate means of the juices of almost any animal reveals the occasional presence of some kind of mobile protozoon, provided with a whip-like process. The enormous group of *Sporozoa (q.v.)* are entirely parasitic, and have been found in every group of animals except the Protozoa and Coelentera. *Infusoria (q.v.)* contain a considerable number of parasitic forms, some endoparasitic; others like the Suctoria, ectoparasitic.

Chapter 7

Plants and Poor Soil

WHEN a soil is sick, either because its beneficial bacteria do not perform their functions properly, or because of abnormalities in its chemical or physical properties, careful treatment and proper cultural methods may restore it to health. But when a soil becomes sick and unproductive because parasitic forms gain a foothold in it, much greater skill and knowledge are required to cope with the problem. Its solution is of the greatest economic importance to the gardener and to the greenhouse man.

Parasitic fungi, upon finding their way into a soil, do not necessarily interfere with the work of the beneficial bacteria, such as the ammonifiers and nitrifiers, for instance. Nor do they always influence the chemical or physical nature of the soil. Many of them directly attack the crop itself, causing serious diseases in the plants.

DAMPING OFF

This disease is very familiar to every grower of plants. It is peculiar to seedlings or tender plants, and is very prevalent in the greenhouse, the hot bed, the cold frame as well as in the field. It is induced by the presence of definite parasitic fungi, which thrive best in overwatered soils, and when the greenhouse is kept at a comparatively high temperature with poor ventilation. Damping off is also favoured by thick sowing and too much shade in the seed bed.

Symptoms of Damping Off. Every experienced grower knows the disease when he sees it. Seedlings freshly damped off are soft and water soaked at the base of the stem. If they are pulled they often break off easily. A more careful examination shows that the root system is entirely decayed, although the upper part of the stem and leaves may still be green, and also possibly fresh.

The degree of prostration in the seedlings is determined by the amount of moisture in the soil. If it is slight, the seedlings will become flabby and wilted before they topple over. With a high moisture content, they are more firm, but become prostrate as soon as infection sets in. The trouble usually begins in spots in the bed, thence spreading in every direction. Damping off is usually caused by several fungi, the chief of which is Pythium de Baryanum Hesse.

The organism was first named and described by Hesse in 1874. Ward found it to be a very prevalent parasite in the garden soils of Europe. In America the fungus was first recognized as of great economic importance by Atkinson. The seedlings of most greenhouse plants may become subject to damping off by Pythium. When examined under a compound microscope, Pythium de Baryanum is seen to be made up of coarse non-separate, highly granular, irregularly branched hyaline vegetative threads or mycelium. The younger growing threads are more finely granular. The oldest are coarsely granular or more often empty. These threads penetrate the cells of the host, where they obtain its food.

Pythium de Baryanum does not often fruit on the dead seedlings. The fruiting is better observed when the fungus is grown in pure culture. Under normal conditions it produces two forms of spores, conidia and oogonia. The summer spores, or conidia, are swellings formed at the tip of the hyphæ. These swellings readily break off from the mother threads and germinate by sending out a slender tube. This tube penetrates the seedling tissue where it grows and develops and after due incubation reproduces the disease.

The oospore, or sexual spore, is the stage which is most commonly found. The female organ (oogonium) first develops as a terminal enlargement which is cut off by a septum from the mother thread. Next or adjacent to it a slender tube is cut off from the mycelium by a septum. This tube performs the function of the male sexual organ and is known as antheridium. The latter then comes into close contact and empties all its content into the female oogonium. Fertilization thus takes place, and a mature egg, or oospore, or winter resting spore is formed.

The latest investigations have not yet disclosed whether or not Pythium de Baryanum is carried over from year to year by its oospores. It is apparently able to propagate itself indefinitely by its vegetative mycelium.

Of the other fungi which are capable of producing a damping off in the greenhouse or seed bed may be mentioned Sclerotinia libertiana Fckl., Phoma solani Halst., Colletotrichum sp., Fusarium sp., Sclerotium rolfsii Sacc. and Rhizactonia solani Kuhn. Each of these, except the last, will be taken up separately in connection with the study of their respective hosts.

The fungi which cause damping off are introduced into the greenhouse, primarily with sick soil used in the compost, and also with infected manure. The practice of dumping diseased plants and all other infected material in the manure pile cannot be too strongly condemned. Sometimes very lightly infected plants with no visible symptoms of disease in the seed bed may nevertheless act as carriers, and like-wise infect the greenhouse soil.

DAMPING OFF OF CUTTINGS

Greenhouse men are often troubled with a damping off of cuttings. In specific cases this is brought about by parasitic fungi which, however, will be taken up at length under the discussion of the various hosts.

ROOT ROT

Although not so virulent as Pythium, Rhizoctonia is a frequent cause of considerable failure in green-house culture. The fungus causes a damping off of seedlings and cuttings and a serious root rot.

Symptoms. The symptoms of Rhizoctonia rot or wilt do not differ materially from those produced by Pythium de

Baryanum. On older plants, however, Rhizoctonia produces cankers or deep lesions which are very characteristic. These are formed on the roots as well as on the base of the stem. The lesions are reddish brown and extend into the cortical or vital layer as well as into the woody tissue.

There is perhaps no other parasitic fungus which is so widespread and which is capable of attacking such a variety of hosts as Rhizoctonia. The work of Peltier shows that the following greenhouse crops are susceptible to Rhizoctonia: beet, bean, cauliflower, celery, cucumber, egg plant, horseradish, lettuce, muskmelon, pepper, radish, tomato, sweet alyssum, amaranthus, ornamental asparagus, china aster, begonia, candytuft, carnation, coleus, dianthus, lavatera, lobelia, pansy, poinsettia, sweet pea, violet.

Cuttings of the following hosts are also reported by Peltier to damp off from Rhizoctonia: Abutilon hybridum, var. lavitzii, Acalypha wilkesiana, var. bicolour, A. wilkesiana, var. tricolour; A. wilkesiana, var. marginata, Ageratum mexicanum, Alyssum odoratum; Coleus, Cuphea phatycentra, Tresine, Petunia, Piguerua trinervia, Lautolina chamoecyparissus, Sedum spectabile, Althernanthera, Vincia major.

The Organism. In the United States the first extended account of Rhizoctonia was given by Pammel. Many other excellent accounts by American workers have appeared from time to time, to which we shall have occasion to refer later.

The genus Rhizoctonia includes several forms of sterile fungi, all of which are distinguished by the manner of growth in pure culture, and by its mycelium. Young hyphae of R. solani Kuhn are at first hyaline, then deepening in colour from a yellowish to a deep brown. The young branches are somewhat narrowed at their point of union with the parent hyphæ and grow in a direction almost parallel with each other. A septum is also laid down at a short distance from the point of union with the parent mycelium.

There is another form of hypha which is made up of barrel shaped cells each of which is capable of germinating like a spore. In pure cultures R. solani produces sclerotia which are first soft, whitish, and which later become hard and dark. The fungus is, carried over from year to year as sclerotia which are able to withstand the effects of heat, cold, drought, or moisture.

Next in importance to Rhizoctonia is a group of fungi which belong to the genus Fusarium. Soils infected with these species of fungi become unfit for tomatoes, sweet peas, etc., thereby causing great financial losses to the greenhouse man. Individual difficulties will be taken up in studying each of these crops separately. As an illustration of a typical Fusarium sick soil let us consider the wilt of sweet pea. The cause of this trouble is a soil inhabiting fungus, Fusarium lath yri Taub.

Symptoms. The first symptom of the disease is a sudden flagging of the leaves, accompanied by general wilting and collapse of the seedling. Usually upon sowing the seeds a fair percentage germinate and reach the height of about 8 to 10 inches before they are attacked by the fungus. If the collapsed seedlings are allowed to remain on the ground, the stems will soon be covered with the sickle shaped spores. Eventually the decayed tissue rots and is soon invaded by small fruit flies which now begin to distribute the fungus from place to place by carrying its spores.

The Organism. The mycelium of Fusarium lath yri is hyaline, septate and branched. At an early age the mycelial cells round up into countless numbers of chlamydospores. Old cultures are practically one mass of these resting bodies. The spores are of two sorts, the macroconidia which are sickle shaped, 3-4 septate, the microconidia are one celled, minute spherical to elliptical.

ROOT KNOT

Caused by Heterodera radicicola (Greef) Mull. Although root knot is most prevalent in light soils, it may sometimes be found in heavier lands. The trouble is most widespread out of doors in the South-ern States, where the winter is mild. In the North the worm is usually unable to winter over in the open unless it is protected by trash or dead weeds. It is, however, prevalent in greenhouses and is undoubtedly introduced with sick soil brought in from the field.

Symptoms. The disease is characterized by swellings or knots on the roots. These swellings may be variously shaped, and are often mistaken for the true nodules of legumes. Infected plants become stunted, pale, and usually linger for a long time before dying.

The Organism. The nematode is a very minute worm, seldom exceeding one twenty-fifth of an inch in length. It is semi-transparent so that it cannot be easily detected by the naked eye. In searching for the eel worm, it is necessary to break a fresh knot. Close examination will reveal two types of worms; a spindle shaped worm, the male, and a pearly white pear shaped organism, the female, firmly embedded in the gall tissue. The female is very prolific, depositing no less than 400 to goo eggs during her lifetime.

The eggs are whitish, semi-transparent, bean shaped bodies, and too small to be noticed without the aid of a magnifying glass. The time which elapses until the eggs hatch depends largely upon weather conditions. In warm days the eggs hatch sooner than in cold days. Upon hatching, the young larvæ either remain in the tissue of the host plant in which they emerge, or, as is more often the case, leave the host and enter the soil. This is the only period during which the worms move about to any great extent in the soil, where they either remain for some length of time or immediately penetrate another root of the host.

The nematodes in most cases become completely buried in the root tissue, establishing themselves in the soft cellular structure which is rich in food. The head of the worm is provided with a boring apparatus consisting of a sharply pointed spear, located in the mouth. This structure not only aids it in getting food but is also valuable in helping the young worms to batter through the cell walls before becoming definitely located. The two sexes during the development are indistinguishable up to fifteen or twenty days, both being spindle shaped. In the molting or shedding of the skin, there is a marked change in the case of the female, especially in the posterior region of the body, which no longer possesses a tail-like appendage. Fertilization occurs soon after this molt, and many radical changes occur in the shape and structure of the organization of the worm.

The fertilized female increases rapidly in breadth and becomes a pearly white flask or pear-shaped individual. At this stage it is far from being worm-like and may, therefore, be overlooked by one unfamiliar with the life-history of the eel worm. The young male is much like that of the young female larvæ, being spindle shaped in outline. The male does not cause as much damage to the root tissue as the female, and its purpose in life seems to be only that of fertilizing the female, for after this function has been performed, it is quite probable that the male worm takes no more food.

Omnivorous Nature of the Eel Worm. There are two hundred and thirty-five species of plants known to suffer from the eel worm. This number includes all the important families of the flowering plants. According to Bessey * the following are among the greenhouse plants subject to root knot: bean, beet, cantaloupe, cauliflower, cucumber, egg-plant, lettuce, radish, tomato. For methods of control.

LEAF BLIGHT NEMATODE

Beside the root knot disease which is caused by Heterodera radicicola there is another nematode which confines its injury to foliage only.

Of the greenhouse hosts affected by this pest may be mentioned the Begonia, Asplenium nidus-avis, Pteris serrulata avistata, Pteris wimeseth, Pteris tremula, and Pelargonium.

Symptoms. On the Cincinnati begonia the symptoms, according to Clinton, are manifested as numerous small indistinct discolourations limited by the small veinlets. In time, however, the tiny spots enlarge and unite, forming a conspicuous reddish-brown blotch. Frequently infection is manifested as long streaks along the rnain veins. Often isolated spots occur in the midst of the surrounding healthy tissue.

On Asplenium nidus-avis, the trouble becomes conspicuous in dark brown areas from the base of the leaf near the midrib. These spread up-ward until the entire lower half of the leaf is killed. On Pteris, the spots appear as reddish brown bands reaching out from the midrib to the border, but limited sidewise by the small parallel cross veins.

The Organism. The nematode in question is a slender microscopical worm. The latter chooses the air chamber of the leaf in which to lay its eggs and upon hatching travels around in different parts of the same leaf or to the neighbouring foliage. The worm can travel only when there is a wet film on the leaves.

Control. Immersing fern plants for five minutes in water heated to 122 degrees F. (50 degrees C.) does not seem to injure the ferns, but seems to kill the nematode. All infected leaves should be cut off and burned. Diseased plants should be isolated from healthy ones. Spraying with Bordeaux may also act as a repellant.

SOIL

Soil, the loose material that covers the land surfaces of Earth and supports the growth of plants. In general, soil is an unconsolidated, or loose, combination of inorganic and organic materials. The inorganic components of soil are principally the products of rocks and minerals that have been gradually broken down by weather, chemical action, and other natural processes. The organic materials are composed of debris from plants and from the decomposition of the many tiny life forms that inhabit the soil.

Soils vary widely from place to place. Many factors determine the chemical composition and physical structure of the soil at any given location. The different kinds of rocks, minerals, and other geologic materials from which the soil originally formed play a role. The kinds of plants or other vegetation that grow in the soil are also important. Topography—that is, whether the terrain is steep, flat, or some combination—is another factor. In some cases, human activity such as farming or building has caused disruption. Soils also differ in colour, texture, chemical makeup, and the kinds of plants they can support.

Soil actually constitutes a living system, combining with air, water, and sunlight to sustain plant life. The essential process of photosynthesis, in which plants convert sunlight into energy, depends on exchanges that take place within the soil. Plants, in turn, serve as a vital part of the food chain for living things, including humans. Without soil there would be no vegetation—no crops for food, no forests, flowers, or grasslands. To a great extent, life on Earth depends on soil.

The study of different soil types and their properties is called soil science or pedology. Soil science plays a key role in agriculture, helping farmers to select and support the crops on their land and to maintain fertile, healthy ground for planting. Understanding soil is also important in engineering and construction. Soil engineers carry out detailed analysis of the soil prior to building roads, houses, industrial and retail complexes, and other structures.

Soil takes a great deal of time to develop—thousands or even millions of years. As such, it is effectively a nonrenewable resource. Yet even now, in many areas of the world, soil is under siege. Deforestation, over-development, and pollution from humanmade chemicals are just a few of the consequences of human activity and carelessness. As the human population grows, its demand for food from crops increases, making soil conservation crucial.

Quality of Soil

The good or bad qualities of a soil have reference to the needs of the crops which are to be grown upon it, and it is only after a consideration of the requirements of plants that a clear conception can be formed of what characters the soil must possess for it to be a suitable medium on which healthy crops can be raised.

In the first place, soil, to be of any use, must be sufficiently loose and porous to allow the roots of plants to grow and extend freely. It may be so compact that root development is checked or stopped altogether, in which case the plant suffers. On the other hand it should not be too open in texture or the roots do not get a proper hold of the ground and are easily disturbed by wind: moreover such soils are liable to blow away, leaving the underground parts exposed to the air and drought.

The roots like all other parts of plants contain protoplasm or living material, which cannot carry on its functions unless it is supplied with an adequate amount of oxygen: hence the necessity for the continuous circulation of fresh air through the soil. If the latter is too compact or has its interstices filled with carbon dioxide gas or with water - as is the case when the ground is water-logged - the roots rapidly die of suffocation just as would an animal under the same conditions.

There is another point which requires attention. Plants need very considerable amounts of water for their nutrition and growth; the waterholding capacity is, therefore, important. If the soil holds too much it becomes water-logged and its temperature falls below the point for healthy growth, at any rate of the kinds of plants. Usually cultivated on farms and in gardens. If it allows of too free drainage drought sets in and the plants, not getting enough water for their needs, become stunted in size. Too much water is bad, and too little is equally injurious.

In addition, the temperature of the soil largely controls the yield of crops which can be obtained from the land. Soil whose temperature remains low, whether from its northerly aspect or from its high water content or other cause, is unsatisfactory, because the germination of seeds and the general life processes of plants cannot go on satisfactorily except at certain temperatures well above freezing-point.

A good soil should be deep to allow of extensive root development and, in the case of arable soils, easy to work with implements. Even when all the conditions above mentioned in regard to texture, water-holding capacity, aeration and temperature are suitably fulfilled the soil may still be barren: plant foodmaterial is needed. This is usually present in abundance although it may not be available to the plant under certain circumstances, or may need to be replenished or increased by additions to the soil of manures or fertilizers.

CHIEF CONSTITUENTS OF THE SOIL

An examination of the soil shows it to be composed of a vast number of small particles of sand, clay, chalk and humus, in which are generally imbedded larger or smaller stones. It will be useful to consider the nature of the four chief constituents just mentioned and their bearing upon the texture, water-holding capacity and other characters which were referred to in the previous section.

Sand consists of grains of quartz or flint, the individual particles of which are large enough to be seen with the unaided eye or readily felt as gritty grains when rubbed between the finger and thumb. When a little soil is shaken up with water in a tumbler the sand particles rapidly fall to the bottom and form a layer which resembles ordinary sand of the seashore or river banks. Chemically pure sand is silicon dioxide (SiO_2) or quartz, a clear transparent glass-like mineral, but as ordinarily met with, it is more or less impure and generally coloured reddish or yellowish by oxide of iron. A soil consisting of sand entirely would be very loose, would have little capacity to retain water, would be liable to become very hot in the daytime and cool at night and would be quite unsuitable for growth of plants.

The term clay is often used by chemists to denote hydrated silicate of alumina ($Al_2O_3.2SiO_2.2H_2O$), of which kaolin or china clay is a fairly pure form. This substance is present in practically all soils but in comparatively small amounts. Even in the soils which farmers speak of as stiff clays it is rarely present to the extent of more than I or 2%. The word " clay " used in the agricultural sense denotes a sticky intractable material which is found to consist of exceedingly fine particles (generally less than. 005 mm. in diameter) of sand and other minerals derived from the decomposition of rocks, with a small amount of silicate of alumina.

The peculiar character which clay possesses is probably due not to its chemical composition but to its physical state. When wet it becomes sticky and almost impossible to move or work with farm implements; neither air nor water can penetrate freely. In a dry state it becomes hard and bakes to a brick. It holds water well and is consequently cold, needing the application of much heat to raise its temperature. It is obvious, therefore, that soil composed entirely of clay is as useless as pure sand so far as the growth of crops upon it is concerned.

Chalk consists, when quite pure, of calcium carbonate (CaC03), a white solid substance useful in small amounts as a plant foodmaterial, though in excess detrimental to growth. Alone, even when broken up into small pieces, it is unsuitable for the growth of plants.

Humus, the remaining constituent of soil, is the term used for the decaying vegetable and animal matter in the soil. A good illustration of it is peat. Its water-holding capacity is great, but it is often acid, and when dr y it is light and incapable of supporting the roots of plants properly. Few of the commonly cultivated crops can live in a soil consisting mainly of humus.

From the above account it will be understood that not one of the four chief soil constituents is in itself of value for the growth of crops, yet when they are mixed, as they usually are in the soils met with in nature, one corrects the deficiencies of the other. A perfect soil would be such a blend of sand, clay, chalk and humus as would contain sufficient clay and humus to prevent drought, enough sand to render it pervious to fresh air and prevent waterlogging, chalk enough to correct the tendency to acidity of the humus present, and would have within it various substances which would serve as food-materials to the crops.

COMPOSITION OF SOILS

Soils comprise a mixture of inorganic and organic components: minerals, air, water, and plant and animal material. Mineral and organic particles generally compose roughly 50 percent of a soil's volume. The other 50 percent consists of pores—open areas of various shapes and sizes. Networks of pores hold water within the soil and also provide a means of water transport. Oxygen and other gases move through pore spaces in soil. Pores also serve as passageways for small animals and provide room for the growth of plant roots.

Inorganic Material

The mineral component of soil is made up of an arrangement of particles that are less than 2.0 mm (0.08in) in diameter. Soil scientists divide soil particles, also known as soil separates, into three main size groups: sand, silt, and clay. According to the classification scheme used by the United States Department of Agriculture (USDA), the size designations are: sand, 0.05 to 2.00 mm (0.002 to 0.08 in); silt 0.002 to 0.05 mm (0.00008 to 0.002 in); and clay, less than 0.002 mm (0.00008 in). Depending upon the rock materials from which they were derived, these assorted mineral particles ultimately release the chemicals on which plants depend for survival, such as potassium, calcium, magnesium, phosphorus, sulfur, iron, and manganese.

Organic Material

Organic materials constitute another essential component of soils. Some of this material comes from the residue of plants, the remains of plant roots deep within the soil, or materials that fall on the ground, such as leaves on a forest floor. These materials become part of a cycle of decomposition and decay, a cycle that provides important nutrients to the soil. In general, soil fertility depends on a high content of organic materials.

Even a small area of soil holds a universe of living things, ranging in size from the fairly large to the microscopic: earthworms, mites, millipedes, centipedes, grubs, termites, lice, springtails, and more. And even a gram of soil might contain as many as a billion microbes—bacteria and fungi too small to be seen with the naked eye.

All these living things form a complex chain: Larger creatures eat organic debris and excrete waste into the soil, predators consume living prey, and microbes feed on the bodies of dead animals. Bacteria and fungi, in particular, digest the complex organic compounds that make up living matter and reduce them to simpler compounds that plants can use for food. A typical example of bacterial action is the formation of ammonia from animal and vegetable proteins.

Other bacteria oxidize the ammonia to form nitrogen compounds called nitrites, and still other bacteria act on the nitrites to form nitrates, another type of nitrogen compound that can be used by plants. Some types of bacteria are able to fix, or extract, nitrogen directly from the air and make it available in the soil.Ultimately, the decay of plant and animal material results in the formation of a dark-coloured organic matter known as humus. Humus, unlike plant residues, is generally resistant to further decomposition.

CHEMICAL COMPOSITION OF THE SOIL

It has been found by experiment that plants need for their nutritive process and their growth, certain chemical elements, namely, carbon, hydrogen, oxygen, nitrogen, sulphur, phosphorus, potassium, magnesium, calcium and iron. With the exception of the carbon and a small proportion of the oxygen and nitrogen, which may be partially derived from the air, these elements are taken from the soil by crops. The following table shows the amounts of the chief constituents removed by certain crops in lb per acre: - Plants also remove from the soil silicon, sodium, chlorine, and other elements which are, nevertheless, found to be unessential for the growth and may therefore be neglected here.

Leguminous crops take some of the nitrogen which they require from the air, but most plants obtain it from the nitrates present in the soil. The sulphur exists in the soil chiefly in the form of sulphates of magnesium, calcium and other metals; the phosphorus mainly as phosphates of calcium, magnesium and iron; the potash, soda and other bases as silicates and nitrates; calcium and magnesium carbonates are also common constituents of many soils. In the ordinary chemical analyses of the soil determinations are made of the nitrogen and various carbonates present as well as of the amount of phosphoric acid, potash, soda, magnesia and other components soluble in strong hydrochloric acid.

Soil Formation

Soil formation is an ongoing process that proceeds through the combined effects of five soil-forming factors: parent material, climate, living organisms, topography, and time. Each combination of the five factors produces a unique type of soil that can be identified by its characteristic layers, called horizons. Soil formation is also known as *pedogenesis* (from the Greek words *pedon*, for "ground," and *genesis*, meaning "birth" or "origin").

Parent Material

The first step in pedogenesis is the formation of parent material from which the soil itself forms. Roughly 99 percent of the world's soils derive from mineral-based parent materials that are the result of weathering, the physical disintegration and chemical decomposition of exposed bedrock. The small percentage of remaining soils derives from organic parent materials, which are the product of environments where organic matter accumulates faster than it decomposes. This accumulation can occur in marshes, bogs, and wetlands.

Bedrock itself does not directly give rise to soil. Rather, the gradual weathering of bedrock, through physical and chemical processes, produces a layer of rock debris called regolith. Further weathering of this debris, leading to increasingly smaller and finer particles, ultimately results in the creation of soil.

In some instances, the weathering of bedrock creates parent materials that remain in one place. In other cases, rock materials are transported far from their source—blown by wind, carried by moving water, and borne inside glaciers.

Climate

Climate directly affects soil formation. Water, ice, wind, heat, and cold cause physical weathering by loosening and breaking up rocks. Water in rock crevices expands when it freezes, causing the rocks to crack. Rocks are worn down by water and wind and ground to bits by the slow movement of glaciers. Climate also determines the speed at which parent materials undergo chemical weathering, a process in which existing minerals are broken down into new mineral components. Chemical weathering is fastest in hot, moist climates and slowest in cold, dry climates.

Climate also influences the developing soil by determining the types of plant growth that occur. Low rainfall or recurring drought often discourage the growth of trees but allow the growth of grass. Soils that develop in cool rainy areas suited to pines and other needle-leaf trees are low in humus.

Living Organisms

As the parent material accumulates, living things gradually gain a foothold in it. The arrival of living organisms marks the beginning of the formation of true soil. Mosses, lichens, and lower plant forms appear first. As they die, their remains add to the developing soil until a thin layer of humus is built up. Animals' waste materials add nutrients that are used by plants. Higher forms of plants are eventually able to establish themselves as more and more humus accumulates. The presence of humus in the upper layers of a soil is important because humus contains large amounts of the elements needed by plants.

Living organisms also contribute to the development of soils in other ways. Plants build soils by catching dust from volcanoes and deserts, and plants' growing roots break up rocks and stir the developing soil. Animals also mix soils by tunneling in them.

WATER IN THE SOIL

The importance of an adequate supply of water to growing crops cannot well be over-estimated. During the life of a plant there is a continuous stream of water passing through it which enters by the root-hairs in the soil and after passing along the stem is given off from the 'stomata of the leaves into the open air above ground. It has been estimated that an acre of cabbage will absorb from the land and transpire from its leaves more than ten tons of water per day when the weather is fine.

In addition to its usefulness in maintaining a turgid state of the young cells without which growth cannot proceed, water is itself a plant food-material and as absorbed from the soil contains dissolved in it all the mineral food constituents needed by plants for healthy nutrition. Without a sufficient supply plants remain stunted and the crop yield is seriously reduced, as we see in dry seasons when the rainfall is much below the average.

If one condition is more necessary than another for good crops it is a suitable supply of water, for no amount of manuring or other treatment of the soil will make up for a deficient rainfall. The amount needed for the most satisfactory nutrition varies with different plants. In the case of fair average farm crops it has been shown that for the production of one ton of dry matter contained in them from 300 to 500 tons of water has been absorbed and utilized by the plants. This may be more than the rainfall, in which case irrigation or special control of the water supply may be necessary.

The water-holding capacity of a soil depends upon the amount of free space between the particles of which it is composed into which water can enter. In most cases this amounts to from 30 to 50% of the volume of the soil.

When the pore-space of the soil is filled with water it becomes water-logged and few plants can effect absorption by their roots under such conditions. The root-hairs die from want of air, and the whole plant soon suffers. Fields of wheat and other cereals rarely recover after a week's submergence, but orchards and many trees when at rest in winter withstand a flooded or water-logged condition of the soil for two or three weeks without damage.

The most satisfactory growth is maintained when the amount of water present is not more than 40 to 60% of what would saturate it. Under such conditions each particle of soil is surrounded by a thin film of water and in the pore-space air can freely circulate. It is from such films that the root-hairs absorb all that plants require for their growth. The movement of water into the root-hairs is brought about by the osmotic action of certain salts in their cell-sap. Crops are, however, unable to absorb all the water present in the soil, for when the films become very thin they are held more firmly or cling with more force to the soil particles and resist the osmotic action of the root-hairs. Plants have been found to wither and die in sandy soils containing i a % of water, and in clay soils in which there was still present 8% of water.

When a long glass tube open at both ends is filled with soil and one end is dipped in a shallow basin of water, the water is found to move upwards through the soil column just as oil will rise in an ordinary lamp wick. By this capillary action water may be transferred to the upper layers of the soil from a depth of several feet below the surface. In this manner plants whose roots descend but a little way in the ground are enabled to draw on deep supplies.

Not only does water move upwards, but it is transferred by capillarity in all directions through the soil. The amount and speed of movement of water by this means, and the distance to which it may be carried, depend largely upon the fineness of the particles composing the soil and the spaces left between each. The ascent of water is most rapid through coarse sands, but the height to which it will rise is comparatively small. In clays whose particles are exceedingly minute the water travels very slowly but may ultimately reach a height of many feet above the level of the " water-table " below.

While this capillary movement of water is of great importance in supplying the needs of plants it has its disadvantages, since water may be transferred to the surface of the soil, where it evapo rates into the air and is lost to the land or the crop growing upon it..

The loss in this manner was found to be in one instance over a pound. of water per day per square foot of surface, the " water-table " being about 4 or 5 ft. below.

One of the most effective means of conserving soil moisture is by " mulching," *i.e.* by covering the surface of the soil with some loosely compacted material such as straw, leaf-refuse or stablemanure. The space between the parts of such substances is too large to admit of capillary action; hence the water conveyed to the surface of the soil is prevented from passing upwards any further except by slow evaporation through the mulching layer. A loose layer of earth spread over the surface of the soil acts in the same way, and a similarly effective mulch may be prepared by hoeing the soil, or stirring it to a depth of one or two inches with harrows or other implements. The hoe and harrow are therefore excellent tools for use in dry weather. Rolling the land is beneficial to young crops. in dry weather, since it promotes capillary action by reducing the soil spaces. It should, however, be followed by a light hoeing or harrowing.

In the semi-arid regions of the United States, Argentina and other countries where the average annual rainfall lies between ioa to 20 in., irrigation is necessary to obtain full crops every year. Good crops, however, can often be grown in such areas without irrigation if attention is paid to the proper circulation of water in the soil and means for retaining it or preventing excessive loss by evaporation. Of course care must be exercised in the selection of plants - such as sorghum, maize, wheat, and alfalfa or lucerne - which are adapted to dry conditions and a warm climate.

So far as the water-supply is concerned - and this is what ultimately determines the yield of crops - the rain which falls upon the soil should be made to enter it and percolate rapidly through its interstices. A deep porous bed in the upper layers is essential, and this should consist of fine particles which lie close to each other without any tendency to stick together and " puddle " after heavy showers. Every effort should be made to prepare a good mealy tilth by suitable ploughing, harrowing and consolidation.

In the operation of ploughing the furrow slice is separated from the soil below, and although in humid soils this layer may be left to settle by degrees, in semi-arid regions this loosened layer becomes. dry if left alone even for a few hours and valuable water evaporates. into the air. To prevent this various implements, such as disk harrows and specially constructed rollers, may be used to consolidate the upper stirred portion of the soil and place it in close capillary relationship with the lower unmoved layer. If the soil is allowed to become dry and pulverized, rain is likely to run off or " puddle the surface without penetrating it more than a very short distance. Constant hoeing or harrowing to maintain a natural soil mulch layer of 2 or 3 in. deep greatly conserves the soil water below. In certain districts where the rainfall is low a crop can only be obtained once every alternate year, the intervening season being devoted to tillage with a view of getting the rain into the soil and retaining it there for the crop in the following year.

Soil scientists also characterize soils according to how effectively they retain and transport water. Once water enters the soil from rain or irrigation, gravity comes into play, causing water to trickle downward. Water is also taken up in great quantities by the roots of plants: Plants use anywhere from 200 to 1,000 kg (440 to 2,200 lb) of water in the formation of 1 kg (2.2 lb) of dry matter. Soils differ in their capacity to retain moisture against the pull exerted by gravity and by plant roots. Coarse soils, such as those consisting of mostly of sand, tend to hold less water than do soils with finer textures, such as those with a greater proportion of clays.

Water also moves through soil pores independently of gravity. This movement can occur via capillary action, in which water molecules move because they are more attracted to the pore walls than to one another. Such movement tends to occur from wetter to drier areas of the soil. The movement from soil to plant roots can also depend on how tightly water molecules are bound to soil particles.

The attraction of water molecules to each other is an example of cohesion. The attraction of water molecules to other materials, such as soil or plant roots, is a type of adhesion. These effects, which determine the so-called matric potential of the soil, depend largely on the size and arrangement of the soil particles. Another factor that can affect water movement is referred to as the osmotic potential. The osmotic potential hinges on the amount of dissolved salts in the soil. Soils high in soluble salt tend to reduce uptake of water by plant roots and seeds. The sum of the matric and osmotic potentials is called the total water potential. In soil, water carries out the essential function of bringing mineral nutrients to plants. But the balance between water and air in the soil can be delicate. An overabundance of water will saturate the soil and fill pore spaces needed for the transport of oxygen.

The resulting oxygen deficiency can kill plants. Fertile soils permit an exchange between plants and the atmosphere, as oxygen diffuses into the soil and is used by roots for respiration. In turn, the resulting carbon dioxide diffuses through pore spaces and returns to the atmosphere.

This exchange is most efficient in soils with a high degree of porosity. For farmers, gardeners, landscapers, and others with a professional interest in soil health, the process of aeration—making holes in the soil surface to permit the exchange of air—is a crucial activity. The burrowing of earthworms and other soil inhabitants provides a natural and beneficial form of aeration.

Bacteria in the Soil.

Recent science has made much progress in the investigation of the micro-organisms of the soil. Whereas the soil used to be looked upon solely as a dead, inert material containing certain chemical substances which serve as food constituents. of the crops grown upon it, it is now known to be a place of habitation for myriads of minute living organisms upon whose activity much of its fertility depends.

They are responsible for many important chemical processes which make the soil constituents more available and better adapted to the nutrition of crops. One cubic centimetre of soil taken within a foot or so from the surface contains from II to 2 millions of bacteria of many different kinds, as well as large numbers of fungi. In the lower depths of the soil the numbers. decrease, few being met with at a depth of 5 or 6 ft.

The efficiency of many substances, such as farm-yard manure,. guanos, bone-meal and all other organic materials, which are spread over or dug or ploughed into the land for the benefit of farm and' garden crops, is bound up with the action of these minute living beings. Without their aid most manures would be useless for plant growth. Farm-yard manure, guanos and other fertilizers. undergo decomposition in the soil and become broken down into compounds of simple chemical composition better suited for absorption by the roots of crops, the changes involved being directly due to the activity of bacteria and fungi. Much of the work carried on by these organisms is not clearly understood; there are, however, certain processes which have been extensively investigated and to these it is necessary to refer.

It has been found by experiment that the nitrogen needed by practically all farm crops except leguminous ones is best supplied in the form of a nitrate; the rapid effect of nitrate of soda when used' as a top dressing to wheat or other plants is well known to farmers.

It has long been known that when organic materials such as the dung and urine of animals, or even the bodies of animals and plants, are applied to the soil, the nitrogen within them becomes oxidized, and ultimately appears in the form of nitrate of lime, potash or some other base. The nitrogen in decaying roots, in the dead stems. and leaves of plants, and in humus generally is sooner or later changed into a nitrate, the change being effected by bacteria. That: the action of living organisms is the cause of the production of nitrates is supported by the fact that the change does not occur when the soil is heated nor when it is treated with disinfectants which destroy or check the growth and life of bacteria.

The process resulting in the formation of nitrates in the soil is spoken of as *nitrification.* The steps in the breaking down of the highly complex nitrogenous proteid compounds contained in the humus of the soil, or applied to the latter by the farmer in the form of dung and organic refuse generally, are many and varied; most frequently the insoluble proteids are changed by various kinds of putrefactive bacteria into soluble proteids (peptones, &c.), these into simpler amido-bodies, and these again sooner or later into compounds 7of ammonia.

The urea in urine is also rapidly converted by the uro-bacteria into ammonium carbonate. The compounds of ammonia thus formed from the complex substances by many varied kinds of micro-organisms are ultimately oxidized into nitrates.

The change takes place in two stages and is effected by two special groups of nitrifying bacteria, which are present in all soils. In the first stage the ammonium compounds are oxidized to nitrites by the agency of very minute motile bacteria belonging to the genus *Nitrosomonas.* The further oxidation of the nitrite to a nitrate is effected by bacteria belonging to the genus *Nitrobacter*. Several conditions must be fulfilled before nitrification can occur. In the first place an adequate temperature is essential; at 5° or 6° C. (40-43° F.) the process is stopped, so that it does not go on in winter. In summer, when the temperature is about 24° C. (75° F.), nitrification proceeds at a rapid rate.

The organisms do not carry on their work in soils deficient in air; hence the process is checked in water-logged soils. The presence of a base such as lime or magnesia (or their carbonates) is also essential, as well as an adequate degree of moisture: in dry soils nitrification ceases.

It is the business of the farmer and gardener to promote the activity of these organisms by good tillage, careful drainage and occasional application of lime to soils which are deficient in this substance. It is only when these conditions are attended to that decay and nitrification of dung, guano, fish-meal, sulphate of ammonia and other manures take place, and the constituents which they contain become available to the crops for whose benefit they have been applied to the land.

Nitrates are very soluble in water and are therefore liable to be washed out of the soil by heavy rain. They are, however, very readily absorbed by growing plants, so that in summer, when nitrification is most active, the nitrates produced are usually made use of by crops before loss by drainage takes place. In winter, however, and in fallows loss takes place in the subsoil water.

There is also another possible source of loss of nitrates through the activity of denitrifying bacteria. These organisms reduce nitrates to nitrites and finally to ammonia and gaseous free nitrogen which escapes into the atmosphere. Many bacteria are known which are capable of denitrification, some of them being abundant in fresh dung and upon old straw. They can, however, only carry on their work extensively under anaerobic conditions, as in waterlogged soils or in those which are badly tilled, so that there is but little loss of nitrates through their agency.

An important group of soil organisms are now known which have the power of using the free nitrogen of the atmosphere for the formation of the complex nitrogenous compounds of which their bodies are largely composed. By their continued action the soil becomes enriched with nitrogenous material which eventually through the nitrification process becomes available to ordinary green crops. This power of " fixing nitrogen," as it is termed, is apparently not possessed by higher green plants.

The bacterium, *Clostridium pasteurianum,* common in most soils, is able to utilize free nitrogen under anaerobic conditions, and an organism known as *Azotobacter chroococcum* and some others closely allied to it, have similar powers which they can exercise under aerobic conditions. For the carrying on of their functions they all need to be supplied with carbohydrates or other carbon compounds which they obtain ordinarily from humus and plant residues in the soil, or possibly in some instances from carbohydrates manufactured by minute green algae with which they live in close union.

Certain bacteria of the nitrogenfixing class enter into association with the roots of green plants, the best-known examples being those which are met with in the nodules upon the roots of clover, peas, beans, sainfoin and other plants belonging to the leguminous order.

That the fertility of land used for the growth of wheat is improved by growing upon it a crop of beans or clover has been long recognized by farmers. The knowledge of the cause, however, is due to modern investigations. When wheat, barley, turnips and similar plants are grown, the soil upon which they are cultivated becomes depleted of its nitrogen; yet after a crop of clover or other leguminous plants the soil is found to be richer in nitrogen than it was before the crop was grown. This is due to the nitrogenous root residues left in the land. Upon the roots of leguminous plants characteristic swollen nodules or tubercles are present.

These are found to contain large numbers of a bacterium termed *Bacillus radicicola* or *Pseudomonas radicicola.* The bacteria, which are present in almost all soils, enter the root-hairs of their host plants and ultimately stimulate the production of an excrescent nodule, in which they live. For a time after entry they multiply, obtaining the nitrogen necessary for their nutrition and growth from the free nitrogen of the air, the carbohydrate required being supplied by the pea or clover plant in whose tissues they make a home.

The nodules increase in size, and analysis shows that they are exceedingly rich in nitrogen up to the time of flowering of the host plant. During this period the bacteria multiply and most of them assume a peculiar thickened or branched form, in which state they are spoken of as bacteroids. Later the nitrogen-content of the nodule decreases, most of the organisms, which are largely composed of proteid material, becoming digested and transformed into soluble nitrogenous compounds which are conducted to the developing roots and seeds. After the decay of the roots some of the unchanged bacteria are left in the soil, where they remain ready to infect a new leguminous crop.

The nitrogen-fixing nodule bacteria can be cultivated on artificial media, and many attempts have been made to utilize them for practical purposes. Pure cultures may be made and after dilution in water or other liquid can be mixed with soil to be ultimately spread over the land which is to be infected.

The method of using them most frequently adopted consists in applying them to the seeds of leguminous plants before sowing, the seed being dipped for a time in a liquid containing the bacteria. In this manner organisms obtained from red clover can be grown and applied to the seed of red clover; and similar inoculation can be arranged for other species, so that an application of the bacteria most suited to the particular crop to be cultivated can be assured. In many cases it has been found that inoculation, whether of the soil or of the seed, has not made any appreciable difference to the growth of the crop, a result no doubt due to the fact that the soil had already contained within it an abundant supply of suitable organisms.

But in other instances greatly increased yields have been obtained where inoculation has been practised. More or less pure cultures of the nitrogen-fixing bacteria belonging to the *Azotobacter* group have been tried and recommended for application to poor land in order to provide a cheap supply of nitrogen.

The application of pure cultures of bacteria for improving the fertility of the land is still in an experimental stage. There is little doubt, however, that in the near future means will be devised to obtain the most efficient work from these minute organisms, either by special artificial cultivation and subsequent application to the soil, or by improved methods of encouraging their healthy growth and activity in the land where they already exist.

Improvement of Soils

The fertility of a soil is dependent upon a number of factors, some of which, such as the addition of fertilizers or manures, increase the stock of available food materials in the soil while others, such as application of clay or humus, chiefly influence the fertility of the land by improving its physical texture.

The chief processes for the improvement of soils are:liming, claying and marling, warping, paring and burning, and green manuring. Most of these more or less directly improve the land by adding to it certain plant food constituents which are lacking, but the effect of each process is in reality very complex. In the majority of cases the good results obtained are more particularly due to the setting free of " dormant " or " latent " food constituents and to the amelioration of the texture of the soil, so that its aeration, drainage, temperature and water-holding capacity are altered for the better.

The material which chemists call calcium carbonate is met with in a comparatively pure state in chalk. It is present in variable amounts in limestones of all kinds, although its white ness may there be masked by the presence of iron oxide and other coloured substances. Carbonate of lime is also a constituent to a greater or lesser extent in almost all soils. In certain sandy soils and in a few stiff clays it may amount to less than 4%, while in others in limestone and chalk districts there may be 50 to 80% present. Pure carbonate of lime when heated loses 44% of its weight, the decrease being due to the loss of carbon dioxide gas.

The resulting white product is termed calcium oxide lime, burnt lime, quicklime, cob lime, or caustic lime. This substance absorbs and combines with water very greedily, at the same time becoming very hot, and falling into a fine dry powder,' calcium hydroxide or slaked lime, which when left in the open slowly combines with the carbon dioxide of the air and becomes calcium carbonate, from which we began.

When recommendations are made about liming land it is necessary to indicate more precisely than is usually done which of the three classes of material named above - chalk, quicklime or slaked lime - is intended. Generally speaking the oxide or quicklime has a more rapid and greater effect in modifying the soil than slaked lime, and this again greater than the carbonate or chalk.

Lime in whatever form it is applied has a many-sided influence in the fertility of the land. It tends to improve the tilth and the capillarity of the soil by binding sands together somewhat and by opening up clays. If applied in too great an amount to light soils and peat land it may do much damage by rendering them too loose and open.

The addition of small quantities of lime, especially in a caustic form, to stiff greasy clays makes them much more porous and pliable. A lump of clay, which if dried would become hard and intractable, crumbles into pieces when dried after adding to it 2% of lime. The lime causes the minute separate particles of clay to flocculate or group themselves together into larger compound grains between which air and water can percolate more freely. It is this power of creating a more crumbly tilth on stiff clays that makes lime so valuable to the farmer. Lime also assists in the decomposition of the organic matter or humus in the soil and promotes nitrification; hence it is of great value after green manuring or where the land contains much humus from the addition of bulky manures such as farm-yard dung.

This tendency to destroy organic matter makes the repeated application of lime a pernicious practice, especially on land which contains little humus to begin with. The more or less dormant nitrogen and other constituents of the humus are made immediately available to the succeeding crop, but the capital of the soil is rapidly reduced, and unless the loss is replaced by the addition of more manures the land may become sterile. Although good crops may follow the application of lime, the latter is not a direct fertilizer or manure and is no substitute for such. Its best use is obtained on land in good condition, but not where the soil is poor.

When used on light dry land it tends to make the land drier, since it destroys the humus which so largely assists in keeping water in the soil. Lime is a base and neutralizes the acid materials present in badly drained meadows and boggy pastures. Weeds, therefore, which need sour conditions for development are checked by liming and the better grasses and clovers are encouraged. It also sets free potash and possibly other useful plant food-constituents of the soil. Liming tends to produce earlier crops and destroys the fungus which causes finger-and-toe or club-root among turnips and cabbages.

Land which contains less than about z /c, of lime usually needs the addition of this material. The particular form in which lime should be applied for the best results depends upon the nature of the soil. In practice the proximity to chalk pits or lime kilns, the cost of the lime and cartage, will determine which is most economical.

Generally speaking light poor lands deficient in organic matter will need the less caustic form or chalk, while quicklime will be most satisfactory on the stiff clays and richer soils. On the stiff soils overl y ing the chalk it was formerly the custom to dig pits through the soil to the rock below. Shafts 20 or 30 ft. deep were then sunk, and the chalk taken from horizontal tunnels was brought to the surface and spread on the land at the rate of about 60 loads per acre. Chalk should be applied in autumn, so that it may be split by the action of frost during the winter.

Quicklime is best applied, perhaps, in spring at the rate of one ton per acre every six or eight years, or in larger doses-

4 to 8 tons - every 15 to 20 years. Small dressings applied at short intervals give the most satisfactory results. The quicklime should be placed in small heaps and covered with soil if possible until it is slacked and the lumps have fallen into powder, after which it may be spread and harrowed in. Experiments have shown that excellent effects can be obtained by applying 5 or 6 cwt. of ground quicklime.

Gas-lime is a product obtained from gasworks where quicklime is used to purify the gas from sulphur compounds and other objectionable materials. It contains a certain amount of unaltered caustic lime and slacked lime, along with sulphates and sulphides of lime, some of which have an evil odour. As some of these sulphur compounds have a poisonous effect on plants, gas-lime cannot be applied to land directly without great risk or rendering it incapable of growing crops of any sort - even weeds - for some time. It should therefore be kept a year or more in heaps in some waste corner and turned over once or twice so that the air can gain access to it and oxidize the poisonous ingredients in it. Many soils of a light sandy or gravelly or peaty nature and liable to drought and looseness of texture can be improved by the addition of large amounts of clay of an ordinary character.

Similarly soils can be improved by applying to them marl, a substance consisting of a mixture of clay with variable proportions of lime. Some of the chalk marls, which are usually of a yellowish or dirty grey colour, contain clay and 50 to 80% of carbonate of lime with a certain proportion of phosphate of lime. Such a material would not only have an influence on the texture of the land but the lime would reduce the sourness of the land and the phosphate of lime supply one of the most valuable of plant foodconstituents. The beneficial effects of marls may also be partially due to the presence in them of available potash. Typical clay-marls are tenacious, soapy clays of yellowish-red or brownish colour and generally contain less than 50% of lime. When dry they crumble into small pieces which can be readily mixed with the soil by ploughing. Many other kinds of marls are described; some are of a sandy nature, others stony or full of the remains of small shells.

The amount and nature of the clay or marl to be added to the soil will depend largely upon the original composition of the latter, the lighter sands and gravel requiring more clay than those of firmer texture. Even stiff soils deficient in lime are greatly improved in fertility by the addition of marls. In some cases as little as 40 loads per acre have been used with benefit, in others 180 loads have not been too much. The material is dug from neighbouring pits or sometimes from the fields which are to be improved, and applied in autumn and winter. When dry and in a crumbly state it is harrowed and spread and finally ploughed in and mixed with the soil.

On some of the strongest land it was formerly the practice to add to and plough into it burnt clay, with the object of making the land work more easily. The burnt clay moreover carried *Cl ay* with it potash and other materials in a state readily available to the crops.

The clay is dug from the land or from ditches or pits and placed in heaps of 60 to ioo loads each, with faggot wood, refuse coals or other fuel. Great care is necessary to prevent the heaps from becoming too hot, in which case the clay becomes baked into hard lumps of brick-like material which cannot be broken up.

With careful management, however, the clay dries and bakes, becoming slowly converted into lumps which readily crumble into a fine powder, in which state it is spread over and worked into the land at the rate of 40 loads per acre.

The paring and burning of land, although formerly practised as an ordinary means of improving the texture and fertility of arable fields, can now only be looked upon as a practice p to be adopted for the purpose of bringing rapidly into cultivation very foul leys or, land covered with a coarse turf. The practice is confined to poorer types of land, such as heaths covered with furze and bracken or fens and clay areas smothered with rank grasses and sedges.

To reduce such land to a fit state for the growth of arable crops is very difficult and slow without resort to paring and burning. The operation consists of paring off the tough sward to a depth of I to 2 in. just sufficient to effectually damage the roots of the plants forming the sward and then, after drying the sods and burning them, spreading the charred material and ashes over the land. The turf is taken off either with the breast plough - a paring tool pushed forward from the breast or thighs by the workman - or with specially constructed paring ploughs or shims. The depth of the sod removed should not be too thick or burning is difficult and too much humus is destroyed unnecessarily, nor should it be too thin or the roots of the herbage are not effectually destroyed.

The operation is best carried out in spring and summer. After being pared off the turf is allowed to dry for a fortnight or so and is then placed in small heaps a yard or two wide at the base, a little straw or wood being put in the middle of each heap, which is then lighted. As burning proceeds more turf is added to the outside of the heaps in such a manner as to allow little access of air. Every care should be taken to burn and char the sod thoroughly without permitting the heap to blaze. The ashes should be spread as soon as possible and covered by a shallow ploughing. The land is then usually sown with some rapidly growing green crop, such as rape, or with turnips.

Paring and burning improves the texture of clay lands, particularly if draining is carried out at the same time. It tends to destroy insects and weeds, and gets rid of acidity of the soil. No operation brings old turf into cultivation so rapidly. Moreover the beneficial effects are seen in the first crop and last for many years. Many of the mineral plant food-constituents locked up in the coarse herbage and in the upper layers of the soil are made immediately available to crops. The chief disadvantage is the loss of nitrogen which it entails, this element being given off into the air in a free gaseous state. It is best adapted for application to clays and fen lands and should not be practised on shallow light sands or gravelly soils, since the humus so necessary for the fertility of such areas is reduced too much and the soil rendered too porous and liable to suffer from drought.

Many thousands of acres of low-lying peaty and sandy land adjoining the tidal rivers which flow into the Humber have been improved by a process termed " warping." The warp consists of fine muddy sediment which is suspended in the tidal river water and appears to be derived from material scoured from the bed of the Humber by the action of the tide and a certain amount of sediment brought down by the tributary streams which join the Humber some distance from its mouth.

The field or area to be warped must lie below the level of the water in the river at high tide. It is first surrounded by an embankment, after which the water from the river is allowed to flow through a properly constructed sluice in its bank, along a drain or ditch to the land which is prepared for warping. By a system of carefully laid channels the water flows gently over the land, and deposits its warp with an even level surface.

At the ebb of the tide the more or less clear water flows back again from the land into the main river with sufficient force to clean out any deposit which may have accumulated in the drain leading to the warped area, thus allowing free access of more warpladen water at the next tide. In this manner poor peats and sands may be covered with a large layer of rock soil capable of growing excellent crops.

The amount of deposit laid over the land reaches a thickness of two or three feet in one season of warping, which is usually practised between March and October, advantage being taken of the spring tides during these months. The new warp is allowed to lie fallow during the winter after being laid out in four-yard " lands " and becomes dry enough to be sown with oats and grass and clover seeds in the following spring. The clover-grass ley is then grazed for a year or two with sheep, after which wheat and potatoes are the chief crops grown on the land.

Green manures are crops which are grown especially for the purpose of ploughing into the land in a green or actively growing state. The crop during its growth obtains a considerable amount of *Green.* carbon from the carbon dioxide of the air, and builds it up *Manuring. i* nto compounds which when ploughed into the land become humus.

The carbon compounds of the latter are of no direct nutritive value to the succeeding crop, but the decaying vegetable tissues very greatly assist in retaining moisture in light sandy soils, and in clay soils also have a beneficial effect in rendering them more open and allowing of better drainage of superfluous water and good circulation of fresh air within them. The ploughing-in of green crops is in many respects like the addition of farm-yard manure.

Their growth makes no new addition of mineral food-constituents to the land, but they bring useful substances from the subsoil nearer to the surface, and after the decay of the buried vegetation these become available to succeeding crops of wheat or other plants. Moreover, where deep-rooting plants are grown the subsoil is aerated and rendered more open and suitable for the development of future crops.

The plants most frequently used are white mustard, rape, buckwheat, spurry, rye, and several kinds of leguminous plants, especially vetches, lupins and serradella. By far the most satisfactory crops as green manures are those of the leguminous class, since they add to the land considerable amounts of the valuable fertilizing constituent, nitrogen, which is obtained from the atmosphere.

By nitrification this substance rapidly becomes available to succeeding crops. On the light, poor sands of Saxony Herr Schultz, of Lupitz, made use of serradella, yellow lupins and vetches as green manures for enriching the land in humus and nitrogen, and found the addition of potash salts and phosphates very profitable for the subsequent growth of potatoes and wheat. He estimated that by using leguminous crops in this manner for the purpose of obtaining cheap nitrogen he reduced the cost of production of wheat more than 50%.

The growing crops should be ploughed in before flowering occurs; they should not be buried deeply, since decay and nitrification take place most rapidly and satisfactorily when there is free access of air to the decaying material. When the crop is luxuriant it is necessary to put a roller over it first, to facilitate proper burial by the plough.

The best time for the operation appears to be late summer and autumn. (J. PE.) *Soil and Disease. - The* influence of different kinds of soil as a factor in the production of disease requires to be considered, in regard not only to the nature and number of the microorganisms they contain, but also to the amount of moisture and air in them and their capacity for heat.

The moisture in soil is derived from two sources—the rain and the ground-water.

Above the level of the ground-water the soil is kept moist by capillary attraction and by evaporation of the water below, by rainfall, and by movements of the ground-water; on the other hand, the upper layers are constantly losing moisture by evaporation from the surface and through vegetation. When the ground-water rises it forces air out of the soil; when it falls again it leaves the soil moist and full of air.

The nature of the soil will largely influence the amount of moisture which it will take up or retain. In regard to water, all soils have two actions - namely, permeability and absorbability. Permeability is practically identical with the speed at which percolation takes place; through clay it is slow, but increases in rapidity through marls, loams, limestones, chalks, coarse gravels and fine sands, reaching a maximum in soil saturated with moisture.

The amount of moisture retained depends mainly upon the absorbability of the soil, and as it depends largely on capillary action it varies with the coarseness or fineness of the pores of the soil, being greater for soils which consist of fine particles. The results of many analyses show that the capacity of soils for moisture increases with the amount of organic substances present; decomposition appears to be most active when the moisture is about 4%, but can continue when it is as low as 2%, while it appears to be retarded by any excess over 4%. Above the level of the ground-water all soils contain air, varying in amount with the degree of looseness of the soil. Some sands contain as much as 50% of air of nearly the same composition as atmospheric air. The oxygen, however, decreases with the depth, while the carbon dioxide increases.

SOIL CHARACTERISTICS

Scientists can learn a lot about a soil's composition and origin by examining various features of the soil. Colour, texture, aggregation, porosity, ion content, and pH are all important soil characteristics.

Colour

Soils come in a wide range of colours—shades of brown, red, orange, yellow, gray, and even blue or green. Colour alone does not affect a soil, but it is often a reliable indicator of other soil properties. In the surface soil horizons, a dark colour usually indicates the presence of organic matter.

Soils with significant organic material content appear dark brown or black. The most common soil hues are in the red-to-yellow range, getting their colour from iron oxide minerals coating soil particles.

Red iron oxides dominate highly weathered soils. Soils frequently saturated by water appear gray, blue, or green because the minerals that give them the red and yellow colours have been leached away.

Texture

A soil's texture depends on its content of the three main mineral components of the soil: sand, silt, and clay. Texture is the relative percentage of each particle size in a soil. Texture differences can affect many other physical and chemical properties and are therefore important in measures such as soil productivity.

Soils with predominantly large particles tend to drain quickly and have lower fertility. Very fine-textured soils may be poorly drained, tend to become waterlogged, and are therefore not well-suited for agriculture. Soils with a medium texture and a relatively even proportion of all particle sizes are most versatile. A combination of 10 to 20 percent clay, along with sand and silt in roughly equal amounts, and a good quantity of organic materials, is considered an ideal mixture for productive soil.

Aggregation

Individual soil particles tend to be bound together into larger units referred to as aggregates or soil peds. Aggregation occurs as a result of complex chemical forces acting on small soil components or when organisms and organic matter in soil act as glue binding particles together.

Soil aggregates form soil structure, defined by the shape, size, and strength of the aggregates. There are three main soil shapes: platelike, in which the aggregates are flat and mostly horizontal; prismlike, meaning greater in vertical than in horizontal dimension; and blocklike, roughly equal in horizontal and vertical dimensions and either angular or rounded. Soil peds range in size from very fine—less than 1 mm (0.04 in)—to very coarse—greater than 10 mm (0.4 in). The measure of strength or grade refers to the stability of the structural unit and is ranked as weak, moderate, or strong. Very young or sandy soils may have no discernible structure.

Porosity

The part of the soil that is not solid is made up of pores of various sizes and shapes—sometimes small and separate, sometimes consisting of continuous tubes. Soil scientists refer to the size, number, and arrangement of these pores as the soil's porosity. Porosity greatly affects water movement and gas exchange.

Well-aggregated soils have numerous pores, which are important for organisms that live in the soil and require water and oxygen to survive. The transport of nutrients and contaminants will also be affected by soil structure and porosity.

Ion Content

Soils also have key chemical characteristics. The surfaces of certain soil particles, particularly the clays, hold groupings of atoms known as ions. These ions carry a negative charge. Like magnets, these negative ions (called anions) attract positive ions (called cations). Cations, including those from calcium, magnesium, and potassium, then become attached to the soil particles, in a process known as cation exchange. The chemical reactions in cation exchange make it possible for calcium and the other elements to be changed into water-soluble forms that plants can use for food. Therefore, a soil's cation exchange capacity is an important measure of its fertility.

pH

Another important chemical measure is soil pH, which refers to the soil's acidity or alkalinity. This property hinges on the concentration of hydrogen ions in solution. A greater concentration of hydrogen results in a lower pH, meaning greater acidity. Scientists consider pure water, with a pH of 7, neutral. The pH of a soil will often determine whether certain plants can be grown successfully. Blueberry plants, require acidic soils with a pH of roughly 4 to 4.5. Alfalfa and many grasses, on the other hand, require a neutral or slightly alkaline soil. In agriculture, farmers add limestone to acid soils to neutralize them.

SOIL USE

For most of human history, soil has not been treated as the valuable and essentially nonrenewable resource that it is. Erosion has devastated soils worldwide as a result of overuse and misuse. In recent years, however, farmers and agricultural experts have become increasingly concerned with soil management.

Erosion

Erosion is the wearing away of material on the surface of the land by wind, water, or gravity. In nature, erosion occurs very slowly, as natural weathering and geologic processes remove rock, parent material, or soil from the land surface. Human activity, on the other hand, greatly increases the rate of erosion. In the United States, the farming of crops accounts for the loss of over 3 billion metric tons of soil each year.

In a cultivated field from which crops have been harvested, the soil is often left bare, without protection from the elements, particularly water. Raindrops smash into the soil, dislodging soil particles. Water then carries these particles away. This movement may take the form of broad overland flows known as sheet erosion. More often, the eroding soil is concentrated into small channels, or rills, producing so-called rill erosion. Gravity intensifies water erosion. Landslides, in which large masses of water-loosened soil slide down an incline, are a particularly extreme example.

Wind erosion occurs where soils are dry, bare, and exposed to winds. Very small soil particles can be suspended in the air and carried away with the wind. Larger particles bounce along the ground in a process called saltation.

Soil Management

To prevent exposure of bare soil, farmers can use techniques such as leaving crop residue in the soil after harvesting or planting temporary growths, such as grasses, to protect the soil from rain between crop-growing seasons. Farmers can also control water runoff by planting crops along the slope of a hill (on the contour) instead of in rows that go up and down.

Soil faces many threats throughout the world. Deforestation, overgrazing by livestock, and agricultural practices that fail to conserve soil are three main causes of accelerated soil loss. Other acts of human carelessness also damage soil.

These include pollution from agricultural pesticides, chemical spills, liquid and solid wastes, and acidification from the fall of acid rain. Loss of green spaces, such as grassland and forested areas, in favour of impermeable surfaces, such as pavement, buildings, and developed land, reduces the amount of soil and increases pressure on what soil remains. Soil is also compacted by heavy machinery and off-road vehicles. Compaction rearranges soil particles, increasing the density of the soil and reducing porosity. Crusts form on compacted soils, preventing water movement into the soil and increasing runoff and erosion.

With the world's population now numbering upwards of 6 billion people—a figure that may rise to 10 billion or more within three decades—humans will depend more than ever on soil for the growth of food crops. Yet the rapidly increasing population, the intensity of agriculture, and the replacement of soil with concrete and buildings all reduce the capacity of the soil to fulfill this need. As a result of an increased awareness of soil's importance, many changes are being made to protect soil. Recent interest in soil conservation holds the promise that humanity will take better care of this precious resource.

POOR SEEDLING EMERGENCE

Patchy crops reduce yield and profit. Poor seedling emergence is more of a problem under minimum tillage, because there is less soil disturbance and more stubble retention. It is less of a problem under conventional tillage, because the seedbeds are finely worked and the seeds are planted into bare ground. However, rather than reject minimum tillage because of its problems, it is preferable to address those problems.

POSSIBLE CAUSES

Poor seedling emergence may be due to:

- Poor seed–soil contact
- Inaccurate seed placement
- A soil temperature that is too low or too high
- Soil insects or soil-borne disease
- Surface crusting after sowing
- Poor quality seed.

Poor seedling vigour and establishment after emergence may caused by the presence of compacted soil beneath the seedling. The soil may have already compacted before sowing, or the planting tine may have smeared the bottom and sides of the seed trench.

Planting Machinery

If poor seedling emergence is due to poor seed–soil contact or inaccurate seed placement, then adjusting your sowing machinery correctly will help to improve emergence.

Surface Crusting

Rain after sowing and before seedling emergence may form a surface crust that reduces emergence. Once seedlings have emerged, the problem of surface crusting is less. However, until the crop reaches full ground cover (and protects the surface from raindrop impact) surface crusting will continue to reduce infiltration.

All soils seal under raindrop impact—even self-mulching soils seal if they are bare. On some soils, the seal is weak or it cracks as it dries; it offers little resistance to seedling emergence. However, on other soils, the seal forms a crust on drying, and this crust prevents seedling emergence. Further light rain will help emergence by softening the surface; further heavy rain will hinder emergence by strengthening the surface crust. Clay soils can crust if the surface is sodic. The treatment for such soils is gypsum.

COMPACTION MANAGEMENT

A small degree of compaction can be beneficial. Soil that is too loose may not be good for plant growth: a loose seedbed dries out too quickly and gives poor contact between the seeds and the soil; seeds may not germinate. The optimum level of compaction depends on the soil types, the plant, and the irrigation management during the growing season. Too much compaction can cause yield reduction. To reduce soil compaction:

- Time mechanical operations carefully
- Reduce axle loads
- Keep livestock off wet cropping country
- Confine traffic to laneways
- Use low ground pressure (wide) tyres (with caution).

The method of compaction repair will depend on the soil's capabilities. A cracking clay that swells and shrinks behaves quite differently to a hardsetting, fine, sandy or silty soil. If a soil swells and shrinks on wetting and drying you can use that natural action to break up a compacted layer. If a hardpan has developed in a soil with little clay, you will need to consider other options.

Example:

- Biological repair—that is, using growing plants to break up a compacted layer or increasing the soil organic matter content

- Mechanical repair—that is, deep ripping or mouldboard ploughing.

BIOLOGICAL REPAIR

Biological repair means using growing plants to restore soil structure. It usually involves a yield penalty. The plants will suffer in restoring the soil structure, and will produce a lower yield. Repair of compaction should therefore work in with the season, the market prices, and your long-term plans.

Biological repair includes:

- biological ripping (using plants to dry and crack the soil)
- Biological drilling (using tap-rooted plants to 'drill' through acompacted layer)
- Increasing soil organic matter content.

Biological Ripping

The easiest way to restore soil structure in compacted cracking clays is to dry the soil, using any plants with a vigourous root system. Drying causes the soil to shrink and crack, allowing water, air and roots to penetrate through the cracks. The process is called biological ripping.

Biological Drilling of Non-Cracking Soils

In soils that don't crack, such as sandy, silty or loamy soils, 'biological ripping' is not an option. However, 'biological drilling' is possible. A strong tap-rooted plant such as lucerne can force its way through a compacted layer and leave root channels for subsequent plant roots to use. This is 'biological drilling'. The compacted layer must be moist: no root will enter dry soil. Depending upon how badly compacted the subsoil is, some roots will not be able to penetrate the compacted layer and the crop will suffer.

Increasing Soil Organic Matter

Some soils, such as fine sandy and silty soils, depend heavily on organic matter for the maintenance of their structures. Even cracking soils benefit from increased organic matter content. Increasing the organic matter of a soil content is a slow process. A pasture phase incorporated into your vegetable crop rotation is perhaps the most costeffective way to achieve this.

MECHANICAL REPAIR

Deep ripping and mouldboard ploughing are sometimes useful for quick results, but they are expensive operations. The initial success of the operation depends on the moisture content of the soil and the depth of tillage. Continuing success depends on managing traffic to avoid having machinery on the soil when it is too wet.

The Importance of Moisture Content

Soil moisture is important when you are working soil. A soil (particularly a clay) that is worked too wet will compact, while a silty or fine sandy soil that is worked too dry may powder. In both instances the soil will then structurally degrade. Before deep ripping, dig a hole to see if there is a compacted layer. If there is, check the layer's depth, thickness, and moisture content.

The severity of compaction will also influence your decision on deep ripping. If there is no compacted layer, there is no need to deep rip. Deep ripping is expensive, and it could do more harm than good if the soil is too wet—a costly way of creating soil structural damage where none existed before! A compacted layer must be at the right moisture content to benefit from the ripping operation When you are cultivating clays (or when the compacted layer is clay) the layer must be dry enough to shatter, rather than smear. When the compacted layer is other than clay, some moisture is desirable so that the layer does not powder.

The depth and thickness of the compacted layer indicates how deep to set the ripper tine. To be effective, the tine should work just below the compacted layer. Dig a hole after a short run and check the effectiveness of the ripping operation.

Chapter 8

Conservation of Forest

CONSERVATION

Conservation, sustainable use and protection of natural resources including plants, animals, mineral deposits, soils, clean water, clean air, and fossil fuels such as coal, petroleum, and natural gas. Natural resources are grouped into two categories, renewable and nonrenewable. A *renewable* resource is one that may be replaced over time by natural processes, such as fish populations or natural vegetation, or is inexhaustible, such as solar energy.

The goal of renewable resource conservation is to ensure that such resources are not consumed faster than they are replaced. *Nonrenewable* resources are those in limited supply that cannot be replaced or can be replaced only over extremely long periods of time. Nonrenewable resources include fossil fuels and mineral deposits, such as iron ore and gold ore. Conservation activities for nonrenewable resources focus on maintaining an adequate supply of these resources well into the future.

Natural resources are conserved for their biological, economic, and recreational values, as well as their natural beauty and importance to local cultures. Tropical rain forests are protected for their important role in both global ecology and the economic livelihood of the local culture; a coral reef may be protected for its recreational value for scuba divers; and a scenic river may be protected for its natural beauty.

Conservation conflicts arise when natural-resource shortages develop in the face of steadily increasing demands from a growing human population. Controversy frequently surrounds how a resource should be used, or allocated, and for whom. A river may supply water for agricultural irrigation, habitat for fish, and water-generated electricity for a factory. Farmers, fishers, and industry leaders vie for unrestricted access to this river, but such freedom could destroy the resource, and conservation methods are necessary to protect the river for future use.

Conflicts worsen when a natural resource crosses political boundaries. The headwaters, or source, of a major river may be located in a different country than the country through which the river flows. There is no guarantee that the river source will be protected to accommodate resource needs downstream. In addition, the way in which one natural resource is managed has a direct effect upon other natural resources. Cutting down a forest near a river, for instance, increases erosion, the wearing away of topsoil, and can lead to flooding. Eroded soil and silt cloud the river and adversely affect many organisms such as fish and important aquatic plants that require clean, clear freshwater for survival.

METHODS OF CONSERVATION

The challenge of conservation is to understand the complex connections among natural resources and balance resource use with protection to ensure an adequate supply for future generations. In order to accomplish this goal, a variety of conservation methods are used. These include reducing consumption of resources; protecting tnem from contamination or pollution; reusing or recycling resources when possible; and fully protecting, or preserving, resources.

Consumption of natural resources rises dramatically every year as the human population increases and standards of living rise. From 1950 to 2000 the world population more than doubled to 6 billion people, with nearly 80 percent living in developing, or poorer, nations. The large, developed nations, however, are responsible for the greatest consumption of natural resources because of their high standards of living. For instance, the average American consumes as much energy as

27 Filipinos or 370 Ethiopians. Conservation education and the thoughtful use of resources is necessary in the developed countries to reduce natural-resource consumption. Reducing the high demand for tropical hardwoods such as teak and mahogany in the United States and Japan would slow the rate of tropical forest destruction.

To protect natural resources from pollution, individuals, industries, and governments have many obligations. These include prohibiting or limiting the use of pesticides and other toxic chemicals, limiting wastewater and airborne pollutants, preventing the production of radioactive materials, and regulating drilling and transportation of petroleum products. Failure to do so results in contaminated air, soil, rivers, plants, and animals. If governments required that all oil tankers be fitted with double-layered hulls, the damages to fisheries and wildlife from the many oil spills of the 20th century, such as the 1967 *Torrey Canyon* oil spill in the English Channel, may have been reduced.

In many cases it is possible to reuse or recycle resources to reduce waste and resource consumption and conserve the energy needed to produce consumer products. Paper, glass, *freon* (a refrigerant gas), aluminum, metal scrap, and motor oil can all be recycled. A preventative measure called *precycling,* a general term for designing more durable, recyclable products such as reusable packaging, encourages reuse. Many states in the United States have established mandatory recycling laws in an attempt to reduce waste and consumption.

Some resources are so unique or valuable that they are protected from activities that would destroy or degrade them. Example, national parks and wilderness areas are protected from logging or mining in the United States because such activities would reduce the economic, recreational, and aesthetic values of the resource.

Forests and wetlands (areas with high soil moisture or surface water) may be protected from development because they enhance air and water quality and provide habitat for a wide variety of plants and animals. Unfortunately, these areas are often threatened with development because it is difficult to measure the economic benefits of cleaner air, cleaner water, and the many other environmental benefits of these ecosystems (the plants and animals of a natural community and their physical environment).

THE ROLES OF FORESTS IN MEETING BASIC HUMAN NEEDS

FOREST CONSERVATION

Forests are influenced by climate, landform and soil composition and they exist in a wide variety of forms in the tropical, temperate and boreal zones of the world. Each forest type, evergreen and deciduous, coniferous and broadleaved, wet and dry, as well as closed and open canopy forests, has its own uniqueness and together these forests complement one another and perform the various socio-economic, ecological, environmental, cultural and spiritual functions.

Recent surveys on a global basis suggest that there are about 1.4 million documented species, and the general consensus is that this is an underestimate - perhaps 5 - 50 million species exist in the natural ecosystems of forests, savannas, pastures and rangelands, deserts, tundra, lakes and seas. Farmers' fields and gardens are also importance repositories of biological resources.

Forests provide many social, economic, and environmental benefits. In addition to timber and paper products, forests provide wildlife habitat and recreational opportunities, prevent soil erosion and flooding, help provide clean air and water, and contain tremendous biodiversity. Forests are also an important defence against global climate change.

Through the process of photosynthesis, forests produce life-giving oxygen and consume huge amounts of carbon dioxide, the atmospheric chemical most responsible for global warming. By decreasing the amount of carbon dioxide in the atmosphere, forests may reduce the effects of global warming.However, huge areas of the richest forests in the world have been cleared for wood fuel, timber products, agriculture, and livestock. These forests are rapidly disappearing. The tropical rain forests of the Brazilian Amazon River basin were cut down at an estimated rate of 14 million hectares (35 million acres) each year—an area about the size of the state of Wisconsin—in the 1990s.

The countries with the most tropical forests tend to be developing and overpopulated nations in the southern hemisphere. Due to poor economies, people resort to clearing the forest and planting crops in order to survive. While there have been effective efforts to stop deforestation directly through boycotts of multinational corporations responsible for exploitative logging, the most effective conservation policies in these countries have been efforts to relieve poverty and expand access to education and health care.

To enhance this networking and to optimise the global representativeness of these biogeographic areas for the conservation of biological diversity, a list of these areas based on mutually agreed terms by national governments should be formulated. It should also include the identification of these biogeographic areas and the development of joint mechanisms, as well as the quantification of the costs involved and the identification of sources of fund needed to manage and conserve these areas.

Joint mechanisms for possible international cooperation to establish transboundary biogeographic areas should also be implemented.

However, it has been recognized that totally protected areas can never be sufficiently extensive to provide for the conservation of all ecological processes and for all species. Nonetheless, there is a need to establish a minimum acceptable national target to be designated as forest conservation areas in each country.

This effort could be further enhanced by establishing buffer zones of natural forests around the protected area where an inner buffer zone is devoted to basic and applied research, environmental monitoring, traditional land use, recreation and tourism or environmental education and training; and an outer buffer zone where research is applied to meet the needs of the local communities. Such management practices are in consonance with Principle 8(e) of the Forest Principles.

Besides the need to set aside conservation areas, it is now being increasingly realised that sustainable production of wood, through environmentally sound selective harvesting practices is one of the most effective ways in ensuring *in-situ* conservation of the biological diversity of forest ecosystems. Such selectively harvested and managed forests will retain most of the diversity of the old-growth/primary forests both in terms of numbers and population of species.

The economic value of the wood and the environmental benefits produced would fully justify investments made in maintaining the forest cover as exemplified in such practices in ensuring its sustainability. The implementation of environmentally sound selective harvesting practices would go a long way in promoting *in-situ* conservation of biological diversity and the sustainable utilisation of the forest resources. In this regard, the establishment of tree plantations would alleviate the pressure on over-harvesting the natural forests in view of the increasing demand of wood from the forests.

The sustainable production of forest goods and services and the conservation of biological diversity in forest ecosystems, as well as the equitable sharing of benefits arising from the utilisation of the genetic resources would require concrete actions at both the national and international levels. In this context, it is imperative that national policy and strategies, among others, should set target on the optimum forest area for forest conservation and for the sustainable production of goods and services, as well as outline relevant measures to enhance both *ex-situ* and *in-situ* forest conservation during forest harvesting. In some cases, long term measures may include the rehabilitation and re-creation of old-growth/primary forests.

In this connection, it is imperative that countries having a high proportion of their land areas under forest cover, especially the developing countries, have access to new and additional financial resources and the "transfer of environmentally sound technologies and corresponding know-how on favourable terms, including on concessional and preferential terms", as reflected in Principles 10 and 11 respectively, of the Forest Principles; in order to ensure the sustainable management, conservation and development of their forest resources.

Moreover, "trade in forest products should be based on non-discriminatory and multilaterally agreed rules and procedures consistent with international trade law and practices" and "unilateral measures,incompatible with international obligations or agreements, to restrict and/or ban international trade in timber or other forest products should be removed or avoided" as called for in Principles 13 (a) and 14 respectively, of the Forest Principles should be respected by the international community, in order to attain long-term sustainable forest conservation and management.

ENHANCEMENT OF FOREST COVER

Enhancement of forest cover is to be viewed as a proactive measure taken to arrest and reverse the current trend of forest decline and degradation.

In this context, the world's forests have been under threat and are declining. It is estimated that forests covered four-fifths of the existing area at the beginning of the Eighteenth century. Of this total, approximately half were in tropical regions and half in temperate and boreal regions.

However, these forests are declining as a result of deforestation. By the mid-Nineteenth century, it was estimated that global forest cover had decreased to 3,900 million hectares or 30 per cent of the world's land area. The latest figure by the Food and Agriculture Organisation of the United Nations as reflected in the Forest Resources Assessment 1990 had estimated that global forest cover as at the end of 1990 had further decreased to 3,188 million hectares or about 24.4 per cent of the world's land area. Processes of reduction and degradation of forest cover have led to an average annual loss of 0.6 per cent.

Although the annual loss of temperate and boreal forests is said to be negligible in recent time, historically, large- scale deforestration had taken place in Europe during the Industrial Revolution to cater for the needs of agricultural expansion, building materials and industrial development (Hinde, 1985). In fact, it is estimated that almost 200 million hectares or more than 50 per cent of the original forest cover had been lost (UN, 1991).

On the other hand, deforestation in the developing world is a rather recent phenomenon due to poverty, indebtedness and the increasing need for food, shelter and energy to cater for the growing population.

In this regard, the four main causes of deforestation in developing countries are shifting cultivation, conversion to agriculture and pasture, wood removals for fuelwood and inappropriate timber utilisation, and the need for infrastructural development. 39.5 per cent of the 1.54 million hectares of closed forest deforested between 1981 and 1990 in Africa was due to agriculture fallow and shifting cultivation, 35.1 per cent due to conversion to mainly permanent agriculture, and the balance 25.4 per cent due to over-exploitation and over-grazing (FAO, 1993a).

However, as a result of improved socio-economic development in Africa, the rate of deforestation due to agriculture fallow and shifting cultivation had in fact decreased by 27.2 per cent when compared to 66.7 per cent which was recorded during the period 1976- 1980 (UN, 1991).

Besides the loss of forest cover through deforestation, there has been a general degradation in the quality and health of global forests due to acid rain and other atmospheric pollutants, especially in developed countries, as well as through forest fires, unsustainable use as a result of inappropriate logging and fuelwood exploitation.

The depletion of global forests and their degradation are causes for concern as they involve not only the loss of forest areas, but also the ultimate quality of the forests. If this trend is unchecked, the implications on the world would be catastrophic. Not only would the existence of all forest types be threatened, but the capability of these forests to perform their various roles and functions in perpetuity would also be seriously undermined. Hence, the need to address the decline in global forest areas and its degradation through enhancing forest cover is immediate.

In this context, is the current global forest cover of 24.4 per cent sufficient? If not, what level of forest cover should we aim for in order to ensure that forest resources and forest lands are sustainably managed to meet the needs of the present and future generations?

At the Ministerial Conference on Atmospheric Pollution and Climate Change held in the Netherlands in November, 1989, the Noordwijk Declaration on Climate Change advocated a world net forest growth of 12 million hectares per year by the turn of the century while a global forest cover of 30 per cent by the year 2000 was proposed at the second Ministerial Conference of Developing Countries on Environment and Development held in Malaysia in April, 1992.

There is every indication that the existing global forest cover should be enhanced through greening of the world. In this connection, restoration of all deforested lands in the industrialised world to close to the original levels of forest coverage is improbable, but this does not mean significant reforestation and afforestation are impossible. All countries which aim for a sound environmental future should set themselves a target of a minimum level of forest cover to be maintained in perpetuity.

Countries having more than 30 per cent of their land areas under forest cover after taking into account their socio-economic development needs, particularly the developing countries, should be given incentives to improve the quality of their forests, as well as assistance given to reduce their dependence on wood especially as fuel. On the other hand, countries having less than 30 per cent of their land areas under forest cover, but have the means must increase and enhance their forest cover through rehabilitation and afforestation, which may include, in some cases, the conversion of heavily subsidised farms back to forests.

As for those countries which are rich but are constrained by physical and climatic conditions to grow trees because of their geographic locations, they could play their roles by assisting the poorer countries in increasing and enhancing their forest cover.

As the future of forests are not only dependent on their quantity, but their quality as well, it is pertinent that all forests, especially those temperate and boreal forests of the developed countries should be protected against air-borne pollutants, particularly that of acid deposition, which are harmful to the health of the forest ecosystems. Appropriate measures should also be taken to protect forests from fire.

ROLES OF FORESTS

A well-managed forest is a constantly self-renewing resource and provides a wide range of benefits at local, national and global levels. Some of these benefits depend on the forest being left untouched or subject to minimal interference while others can only be realised by harvesting the forest. Among the most important roles of forests are sustainable production of wood and timber products, provision of food, shelter and energy, mitigation of climate change, conservation of water and soil, as well as for recreation and ecotourism. Forests are also important repositories of biological diversity.

In this regard, wood is of major economic importance as in 1990 the world's production of industrial timber was about 1,600 million cubic metres, of which about 75 per cent came from the developed countries, while international trade in wood and wood products, as well as paper and pulp is estimated to worth US$96,000 million a year, of which about US$12,500 million comes from developing country exports (FAO, 1993b). Besides, currently fuelwood comprises about 85 per cent of the wood consumed in the developing countries and accounts for more than 75 per cent of total energy consumption in the poorest countries and that over 2,000 million people use fuelwood as the primary source of fuel (UN, 1991).

In recent years, attention has also been focused on the importance of non-wood forest products which include plants for food and medicinal purposes, fibres, dyes, animal fodder and other necessities. Indonesia, earns an estimated US$120 million a year from rattans, resins, sandalwood, honey, natural silk and pharmaceutical and cosmetic compounds (FAO 1990), while the local production of bidi cigarette from the tendu leaf *(Diospyros melanoxylon)* in India provides part-time employment for up to half a million women (FAO, 1993b). In this connection, it has been estimated that more than 200 million people in the tropics live in the forests (FAO, 1993b) and in some parts of Africa as much as 70 per cent of animal protein comes from forest games such as birds and rodents (FAO, 1990).

The economic value of forests in relation to floods and soil conservation is that they may allow for agricultural and even industrial development on floodplains because they contribute to the mitigation of the effects of floods and in minimizing soil erosion especially in mountainous and hilly areas. In fact a well- managed forest would provide a number of goods and services to meet basic human needs as outlined in Annex I.

FOREST CONSERVATION

- To strengthen efforts in forest conservation and the sustainable management of forest resources, it is imperative to ensure the participation of local community and that all national policy and strategies must indicate the forest area set aside for forest conservation and in the sustainable production of forest goods and services. In this context, developing countries must have access to new and additional financial resources and the transfer of environmentally sound technologies.
- To further ensure sustainable forest conservation and sustainable forest management, the prices of timber and timber products at the market place must fully reflect both their replacement and environmental costs, and that trade in forest products should be non-discriminatory and any unilateral measures to restrict and/or ban their trade should be removed or avoided. Moreover, expenses needed for sustainable forest

management, including reforestation and afforestation must be included into the cost of all kinds of production obtained from the forest resources.

- A global network of well-managed and adequately funded protected areas be established. In this regard, a list of biogeographic areas that is mutually agreed by national governments should be prepared to ensure global representativeness of forest conservation areas.
- In order to ensure the sharing on mutually agreed terms of benefits and profits, including biotechnology products derived from the utilisation of biological diversity, efficient and cost-effective methodologies should be developed to assess the biological resources of forests at the genetic, species and ecosystem levels, including the development of techniques to ascribe economic values to these resources.
- In the light of the agreement at UNCED and in accordance with the requirements of the Convention on Biological Diversity, existing forest harvesting practices should be critically reviewed to ensure effective *in-situ* conservation of biological diversity during forest utilisation. Countries should also endeavour to identify forest ecosystems or even landscapes that are threatened with irreversible changes, as well as their causes so as to enable prompt actions to be taken to arrest them.

ENHANCEMENT OF FOREST COVER

- Maintaining and enhancing forest cover, reforestation or afforestation will incur costs, either from opportunities foregone for alternative uses, or from benefits lost from existing land uses. Policy responses must take this into account. The legitimate rights of countries over their natural resources must be upheld. An equitable framework must be found to provide adequate compensation to those countries who undertake action to sustainably manage their forests in the wider interests of global environmental enhancement.

- All countries should work towards increasing their level of forest cover to be achieved over a speci- fied time-frame and actions be taken to prepare and implement national forestry action programmes and/ or plans for the management, conservation and sustainable development of forests as called for in para 11.12(b). Countries having less than 30 per cent of their land areas under forest cover, but have the means must undertake concerted efforts to increase their forest cover while rich countries which are constrained by physical and climatic factors to increase their forest cover could assist the poorer nations in increasing and enhancing their forest cover. Countries having more than 30 per cent of their land areas under forest cover after taking into account their socio-economic development needs should be recognised and appropriate incentives should be given to encourage them to improve the quality of their forests.

ROLES OF FORESTS

- To effectively enhance the roles of forests in meeting basic human needs, it is extremely important that the underlying causes of deforestation such as poverty, population pressures, the need for food, shelter and fuel, as well as indebtedness, particularly in the developing countries, must be critically addressed. A consultative and participatory approach should be adopted involving all stakeholders.
- For the development of management measures to be effective, full knowledge on the distribution and values of non-wood forest resources should be made available at the level compatible to those currently available for the wood resources.
- At the landscape level, each territory should set a minimum area of forest land to safeguard the climate-

and-water characteristics of the forest and that the integrity of the forest ecosystem is protected.

- Public awareness of the roles of forests should be strengthened at the level of social and professional groups, as well as at the family level so as to ensure that the important ecological and environmental functions of forests are further enhanced for both the present and future generations.

CURRENT TYPES OF CONSERVATION ISSUES

There are a variety of basic conservation methods used to protect global natural resources. Although each resource has a unique set of conservation problems and solutions, all resources are interconnected in a complex and little-understood web. Scientists have learned that damaging one thread of the web may weaken the entire structure. It is important that this connectivity be addressed in the search for solutions to resource shortages. It would be impractical to work toward the conservation of soil, for instance, without considering the needs and effects of nearby water and vegetation resources.

BIODIVERSITY CONSERVATION

Biodiversity, or biological diversity, denotes the number and variety of different organisms and ecosystems in a certain area. Preserving biodiversity is essential for ecosystems to respond flexibly to damage or change. A single-species corn crop may be quickly destroyed by a certain insect or disease, but if several different species of corn are planted in the field, some of them may resist the insect or disease and survive.

The same principle applies to natural areas, which adapt to natural environmental changes such as wildfire, drought, or disease because of the biodiversity that has evolved in the area over thousands, or even millions, of years. Many forests, such as those that burned in the 1988 fires in Yellowstone National Park in Wyoming, are able to quickly regenerate because many of the plants that thrive there have adapted to fire. Some trees, such as lodgepole pine, may even require fire to aid in reproduction. These trees produce cones that are opened by extreme heat. The fire opens the cones and the seeds are then released into the soil.

Humans benefit greatly from the many medicines, crops, and other products that biodiversity provides. As many as 40 percent of our modern pharmaceutical medicines are derived from plants or animals.

For instance, a small plant from Madagascar, the rosy periwinkle, produces substances that are effective in fighting two deadly cancers, Hodgkin's disease and leukemia.

Unfortunately, human activities have greatly reduced biodiversity around the world. The 20th century encompasses one of the greatest waves of extinction, or elimination of species, to occur on the planet. The greatest threat to biodiversity is loss of habitat as humans develop land for agriculture, grazing livestock, industry, and habitation. The most drastic damage has occurred in the tropical rain forests, which cover less than seven percent of the Earth's surface but contain well over half of the planet's biodiversity. Only 8 percent of the rain forests in Madagascar, home of the rosy periwinkle, remain intact.

Several nations have laws protecting endangered species. An international treaty, the Convention on International Trade in Endangered Species of Wild Fauna and Flora (CITES), went into effect in 1975 and outlawed trade of endangered animals and animal parts. In the United States, the Endangered Species Act (ESA) was enacted in 1973 to protect endangered or threatened species and their habitats. A new scientific field, conservation biology, studies ways to stop the destruction of biodiversity and restore natural habitats.

SOIL CONSERVATION

Soil, a mixture of mineral, plant, and animal materials, is essential for most plant growth and is the basic resource for agricultural production. Soil-forming processes may take thousands of years, and are slowed by natural erosion forces such as wind and rain. Humans have accelerated these erosion processes by developing the land and clearing away the vegetation that holds water and soil in place. The rapid deforestation taking place in the tropics is especially damaging because the thin layer of soil that remains is fragile and quickly washes away when exposed to the heavy tropical rains. Globally, agriculture accounts for 28 percent of the nearly 2 billion hectares (5 billion acres) of soil that have been degraded by human activities; overgrazing is responsible for 34 percent; and deforestation is responsible for 29 percent.

In addition to reducing deforestation and overgrazing, soil conservation involves reforming agricultural soil management methods. Some of the most effective methods include strip-cropping, alternating strips of crop and uncultivated land to minimize erosion and water runoff; contour farming, planting crops along the contours of sloping lands to minimize erosion and runoff; terracing, which also reduces erosion and runoff on slopes; growing legumes, such as clover or soybeans, to restore essential nitrogen in the so and minimizing tillage, or plowing, to reduce erosion.

WATER CONSERVATION

Clean freshwater resources are essential for drinking, bathing, cooking, irrigation, industry, and for plant and animal survival. Unfortunately, the global supply of freshwater is distributed unevenly. Chronic water shortages exist in most of Africa and drought is common over much of the globe. The sources of most freshwater supplies—groundwater (water located below the soil surface), reservoirs, and rivers—are under severe and increasing environmental stress because of overuse, water pollution, and ecosystem degradation. Over 95 percent of urban sewage in developing countries is discharged untreated into surface waters such as rivers and harbors.

About 65 percent of the global freshwater supply is used in agriculture and 25 percent is used in industry. Freshwater conservation therefore requires a reduction in wasteful practices like inefficient irrigation, reforms in agriculture and industry, and strict pollution controls worldwide.

In addition, water supplies can be increased through effective management of *watersheds* (areas that drain into one shared waterway). By restoring natural vegetation to forests or fields, communities can increase the storage and filtering capacity of these watersheds and minimize wasteful flooding and erosion. Restoration and protection of wetlands is crucial to water conservation. Like giant sponges, wetlands stabilize groundwater supplies by holding rainfall and discharging the water slowly, acting as natural flood-control reservoirs.

ENERGY CONSERVATION

All human cultures require the production and use of *energy*—that is, resources with the capacity to produce work or power. Energy is used for transportation, heating, cooling, cooking, lighting, and industrial production. The world energy supply depends on many different resources including traditional fuels such as firewood and animal waste, which are significant energy sources in many developing countries. Fossil fuels account for more than 90 percent of global energy production but are considered problematic resources.

They are nonrenewable—that is, they can be depleted, and their use causes air pollution. In particular, coal plants have been one of the worst industrial polluters since the beginning of the Industrial Revolution of the 19th century. Moreover, mining or drilling for fossil fuels has caused extensive environmental damage.

There is a global need to increase energy conservation and the use of renewable energy resources. Renewable alternatives such as waterpower (using the energy of moving water, such as rivers), solar energy (using the energy from the sun), wind energy (using the energy of the wind or air currents), and geothermal energy (using energy contained in hot-water deposits within the Earth's crust) are efficient and practical but largely underutilized because of the ready availability of inexpensive, nonrenewable fossil fuels in industrial countries.

While some countries, such as France and Japan, depend heavily on nuclear energy (energy produced by atomic fission, or splitting of the atom), it is still not a major energy source. Excessive production costs, serious safety concerns, and problems with the handling of the dangerous radioactive wastes have virtually eliminated it as a viable energy source in the United States.

In addition to using alternative energy resources such as solar and wind power, energy conservation measures include improving energy efficiency. For instance, transportation accounts for most of the oil consumption in the United States. Encouraging the expansion and use of public transportation systems and carpooling dramatically increases energy efficiency. In the household, energy can be conserved by turning down thermostats, switching off unnecessary lights, insulating homes, and using less hot water.

HISTORY OF CONSERVATION

Until the advent and spread of Christianity and Islam in the 4th and 5th centuries, there were many religions based on animism, the belief that all objects have a spiritual being. This belief led to careful stewardship, or protection, of natural resources out of fear or respect for these spiritual beings. Moreover, early agricultural lifestyles, dependent on nature to provide good crops and growing conditions, also encouraged sound land-use practices.

Ancient Phoenicians, Greeks, and Romans developed irrigation, crop rotation, and terraced hillsides as early methods of water, nutrient, and soil conservation.

In Europe, the relationship between humanity and nature became strained with the beginning of the Industrial Revolution in the 19th and early 20th centuries. Industrialization stifled traditional agricultural lifestyles and encouraged urbanization and the marriage of science and technology to control nature and extract resources.

The Industrial Revolution led to environmental damage on a grand scale as European technology spread around the globe. Coal-burning and iron-smelting produced waste that contaminated air and water, the concentrated populations in urban areas produced huge amounts of unconfined raw sewage that contaminated drinking water, and vast forests and plains were cleared for agriculture. The modern conservation movement of the United States began in the mid-19th century when resource depletion and pollution were first becoming serious problems. Westward expansion was encouraged by the government—the Homestead Act of 1862 provided free land to settlers willing to clear it.

Because land ownership required land-clearing, the rapid migration often resulted in barren landscapes. The extensive land-clearing and the rapid depletion of wildlife resources such as buffalo and beaver heralded a public outcry. This concern was reflected in the writings of public figures such as American essayist Ralph Waldo Emerson and naturalist author Henry David Thoreau.

As conservation ideas gained support, a wave of conservation activity swept the country. The world's first national park, Yellowstone National Park, was established in Wyoming in 1872 to protect an area of incredible natural beauty. In 1873, the American Association for the Advancement of Science petitioned Congress to halt unwise use of natural resources, the Forest Reserve Act of 1891 authorized what would become known as National Forests, and the Lacey Act of 1900 established the first wildlife protection measures by restricting commercial hunting and the trade of illegally killed animals.

The administration of President Theodore Roosevelt (1901-1909) was noted for its conservation achievements. Roosevelt set aside a total of almost 94 million hectares (235 million acres) of public lands to protect them from exploitation by private interests. He installed forestry expert Gifford Pinchot as the head of the new U.S. Forest Service in 1905 and adopted Pinchot's principle of *multiple use,* the nation's first formal natural-resource policy. The multiple-use policy advocated scientific management of public lands for a variety of uses, including commercial development.

This conservation policy was not popular among many Americans who backed full preservation of natural areas. Naturalist and author John Muir believed that any commercial development of natural areas was inappropriate. A powerful rift soon developed between multiple-use advocates and preservationists. This rift came to a climax during the 12-year battle over a plan to dam the Tuolumne River in Hetch Hetchy Valley in California, and the controversy still exists today.

A renewed surge of public conservation activity occurred during the Great Depression of the 1930s. In an attempt to encourage conservation and stimulate the economy, President Franklin D. Roosevelt established the Civilian Conservation Corp in 1933, which provided two million jobs planting trees, building dams and irrigation systems, and establishing soil conservation and wildlife protection programs.

The conservation movement rose into the spotlight again in the 1960s as publications such as *Silent Spring* (1962) by American biologist Rachel Carson raised public concerns about the health and environmental hazards of pesticides and other toxic chemicals used by industry. Several catastrophic events in 1969, including the toxic waste fires on the Cuyahoga River in Cleveland, Ohio and a coastal oil spill in Santa Barbara, California focused media attention on the need for environmental conservation.

The estimated 20 million people across the United States who attended the first national Earth Day, a day for recognizing environmental concerns, on April 22, 1970, demonstrated massive public support for conservation issues. Conservation legislation passed in the 1970s included the Endangered Species Act, the Marine Mammal Protection Act, the Clean Air Act, the Clean Water Act, and the Toxic Substance Control Act.

The 1980s experienced a slowdown of the conservation momentum of the 1970s. Resource conservation concerns remained in the public mind, however, due to continued scientific discoveries concerning global warming, acid rain (a harmful mix of precipitation and damaging pollutants), and depletion of the ozone layer (a gaseous layer in the atmosphere that protects Earth from the sun's harmful ultraviolet rays). Ecological disasters such as the nuclear reactor explosion near the Ukrainian town of Chernobyl' in 1986 and the tanker *Exxon Valdez* oil spill in Prince William Sound, Alaska in 1989, served as catastrophic reminders of the effects of human carelessness.

The European conservation movement began to grow as the effects of industrialization worsened in the mid-20th century. Clean air legislation was enacted in the United Kingdom in 1956 in reaction to London's industrial smog, which killed more than 2,000 people in early December 1952. Political parties with environmental or conservation agendas sprang up in New Zealand, Australia, and Europe by the 1970s, and became known as Green Parties in the 1980s. In the 1970s and 1980s, courageous grassroots organizations such as the Chipko movement in India (a coalition of villagers, mostly women) and the Brazilian rubber tappers (workers who extract chicle, the tree sap used to make rubber) fought for preservation of the forests that provided their livelihood.

In 1972 the United Nations Environment Program was formed to encourage international cooperation in conservation and development strategies. Collaboration on environmental conservation issues included the 1987 Montreal Protocol to protect the ozone layer, the 1992 United Nations Conference on Environment and Development (UNCED) in Rio de Janeiro, Brazil, and the 1994 United Nations Conference on Population and Development in Cairo, Egypt. The participation of the United States in this international movement was weak, while Canadian and European support and participation were strong.

The 1992 UNCED Conference, commonly referred to as the Earth Summit, was the largest international meeting ever held with 178 nations participating.

Its proceedings noted the economic and environmental gulf between the northern and southern hemispheres and emphasized a sustainable growth, utilitarian approach to conservation. In the same year an appeal entitled *World Scientists' Warning to Humanity* was released.

This paper was signed by 1,700 of the world's leading scientists (including 104 Nobel laureate scientists), 19 national academies of science, and the director general of the United Nations Educational, Scientific, and Cultural Organization (UNESCO). It stated that at the current rate of consumption, the Earth's resources may soon be reduced to the point at which the living world would be "unable to sustain life in the manner that we know." In several publications, ecologists and economists agree that despite the immediate benefits of economic growth, infinite growth in material and energy consumption is not compatible with the finite resources of the Earth and will undermine the well-being of both economic and ecological systems. For these reasons, natural resource conservation has become one of the most important challenges to face the human race.

FORESTRY

Forestry, management of forestlands for maximum sustained yield of forest resources and benefits. Although forestry originally concerned mainly timber production, it now also involves the management of grazing areas for domestic livestock, the preservation of wildlife habitats, watershed protection, and the development of recreational opportunities. The management of forestlands therefore helps to ensure that wooded areas are used for maximum benefit according to their nature.

Attempts to regulate and manage forestlands occurred well before the 19th century in Rome, the Middle East, China, and Western Europe. Wars and political instability invariably interfered with these efforts; so destructive use of forests was common in the more densely populated countries. Early settlers in the U.S., regarded forests as impediments to cultivation and sought to remove them as quickly as possible.

This attitude fostered a "cut and get out" philosophy among timber operators, and exploitative logging persisted well into the 20th century. Scientific forestry was initiated in the U.S. at the end of the 19th century largely through the influence of the federal government, with measures including the establishment of the Division of Forestry in 1885 and authorization of forest reserves in 1897. The latter became known as the National Forests in 1905, when the jurisdiction of reserves was transferred from the Department of the Interior to the Department of Agriculture, and the Division of Forestry became the Forest Service. Since that time, forest management has been practiced on state and private lands as well as on federal holdings.

Education in technical forestry began in Western Europe about 1825 with private schools in France and Germany; these were later replaced by state institutions. Until World War II, Western Europe was the world leader in scientific forestry, and many famous schools and institutes were established there. In the U.S. the first forestry curriculum was established in 1898. Forestry education is now offered in many universities throughout the country and elsewhere in the world.

Forestry comprises such specialties as dendrology, silviculture, forest protection, mensuration, engineering, utilization, and management. Dendrology concerns tree identification, distribution, age determination, and species characteristics. Silviculture studies the relationship of a forest to its environment and involves the development, care, and reproduction of stands of timber. Forest protection examines sources of injury, including disease, erosion, insect and animal destruction, and fire.

The science of measuring forest stands, including rate of growth and potential yield of standing timber, is called mensuration. Forest engineering concerns the mechanics of cutting and transportation involved in modern timber growing and harvesting.

Forest utilization considers the properties of wood and its conversion into lumber and wood products. Forest management applies business methods and the principles of technical forestry to the overall operation of forestlands.

The practice of technical forestry includes many operations, from tree planting to harvesting. Central to the operation is the cycle of cutting and replenishment. Four major methods have been developed for this purpose: clear-cutting, selection, seed tree, and shelterwood. In clear-cutting all trees in a given area are cut, and reproduction is obtained by artificial planting or by natural seeding from trees bordering the cleared areas.

This method, adapted to light-demanding species, produces even-aged stands, allows control of stand composition, and is conducive to mechanized harvesting and disposal of slash, or logging debris. Selection cutting maintains a forest of mixed ages from which the largest and most mature trees are harvested periodically.

Although this method is expensive and may cause injury to younger trees during logging, it provides continuous cover and an attractive forest and involves sounder ecological practices. In the seed tree method, about 10 percent of the trees in the cutting area are left in an evenly spaced pattern as a natural seed source. The shelterwood method, which involves the removal of the mature trees in cuttings over a 10- to 15-year period, promotes natural reproduction and produces relatively even-aged stands.

Other practices contribute to the development of commercial forestry. These include artificial planting by direct seeding or transplants, especially in conjunction with clear-cutting; treatment with fertilizers to increase production; and the selection and breeding of timber trees, producing strains that excel in growth, disease resistance, or other desirable characteristics. These measures fit into the pattern of an intensive forestry in which greater production is obtained through technical knowledge and increased mechanization.

Chapter 9

Deforestation

Environmental issues effect every life on this planet from the smallest parasite to the human race. The reason for this is simple. A single disruption in the Earthÿÿs delicate balance can mean certain destruction of the very place that cradles the lives of many species. What is not so simple is finding alternatives to the now dangerous and confronting acts of planet degradation that have been afflicted on the planet over recent years. One such issue that requires consideration is deforestation. Trees have been or are being cut down at increasingly high rates. If this is not stopped many unfavorable side effects could result.

Deforestation, indiscriminate cutting or over-harvesting of trees for lumber or pulp, or to clear the land for agriculture, ranching, construction, or other human activities. Forests cover around a quarter to a third of the total land surface of the Earth. The reduction in area of this valuable environmental, social and economic resource through deforestation has the potential to cause problems on a global scale. Climate models have demonstrated a clear link between deforestation and climate change. Deforestation occurs when areas that used to be forested are converted for use in other ways such as logging or as agricultural ground. Deforestation can be at the heart of many problems that affect our environment.

RAIN FOREST

Rain Forest, woodland characterized by lush vegetation and comparatively high temperature and rainfall throughout the year. Rain forests are the world's most biologically diverse ecosystems. Although they account for less than 7 percent of the land surface on Earth, they contain more than 50 percent—some scientists estimate as high as 90 percent—of its plant and animal species. One hectare (about 2.5 acres) of tropical rain forest may contain more than 600 species of trees. By comparison, the forests of the United States and Canada combined contain only around 700 tree species.

Even more impressive are the number and diversity of animal species that call rain forests their home. One study found more species of ants living on a single rain forest stump than exist in all of the British Isles.

Rain forests also play a critical role in global climate regulation by absorbing carbon dioxide, a greenhouse gas believed to be principally responsible for global warming. Plants naturally absorb carbon dioxide and give off oxygen gas in the process of photosynthesis, and tropical rain forests absorb more carbon dioxide than any other terrestrial ecosystem on earth.

Global emissions of carbon dioxide have increased nearly 30 percent in the last century. There is general agreement among the scientific community that by absorbing some of the gas, rain forests play a vital role in lessening its impacts.

To be classified as a rain forest, a forest must have a *closed canopy*, in which the treetops, or crowns, touch each other, creating a shaded forest interior. In addition, temperature and rainfall must be high and relatively even throughout the year. Forests that meet these criteria are found flanking the equator in South and Central America, Asia, Africa, and Australia. In South America, a vast, forested area of the Amazon River basin in Brazil and neighbouring countries is by far the largest rain forest in the world.

It encompasses more than 3.5 million sq km (about 1.4 million sq mi)—about half of the total global rain forest cover. The larger of two large rain forests in Asia is centreed along the Malay Archipelago, including the islands of Borneo and Sumatra, the Malay Peninsula, and the Republic of the Philippines. The other main rain forest in Asia is found primarily on the island of New Guinea and in northern

Australia. In Africa, most of the rain forest is concentrated along the Atlantic coast and the Congo River Basin.

In regions of the northern and southern hemispheres, small areas of temperate rain forest are found along the coasts where rainfall and humidity are high and winters are mild. Such temperate rain forests are found in the Pacific Northwest of the United States and southwestern Canada. They are dominated by only a few tree species, however, making them differ dramatically from highly diverse tropical rain forests. This article focuses on tropical rain forests.

THE CAUSES OF RAINFOREST DESTRUCTION

Immediate Causes

The immediate causes of rainforest destruction are clear. The main causes of total clearance are agriculture and in drier areas, fuelwood collection. The main cause of forest degradation is logging. Mining, industrial development and large dams also have a serious impact. Tourism is becoming a larger threat to the forests.

Logging

Commercial logging companies cut down mature trees that have been selected for their timber. The timber trade defends itself by saying that this method of 'selective' logging ensures that the forest regrows naturally and in time, is once again ready for their 'safe' logging practices (WWF).

In most cases, this is untrue due to the nature of rainforests and of logging practices. Large areas of rainforest are destroyed in order to remove only a few logs. The heavy machinery used to penetrate the forests and build roads causes extensive damage. Trees are felled and soil is compacted by heavy machinery, decreasing the forest's chance for regeneration.

The felling of one 'selected' tree, tears down with it climbers, vines, epiphytes and lianas. A large hole is left in the canopy and complete regeneration takes hundreds of years.

Removing a felled tree from the forest causes even further destruction, especially when it is carried out carelessly. It is believed that in many South East Asian countries 'between 45-74% of trees remaining after logging have been substantially damaged or destroyed' (WWF).

The tracks made by heavy machinery and the clearings left behind by loggers are sites of extreme soil disturbance which begin to erode in heavy rain. This causes siltation of the forests, rivers and streams. The lives and life support systems of indigenous people are disrupted as is the habitat of hundreds of birds and animals.

Little if any industrial logging of tropical forests is sustainable. The International Tropical Timber Organisation (ITTO), the body established to regulate the international trade in tropical timber, found in 1988 that the amount of sustainable logging was "on a world scale, negligible".

"Logging roads are used by landless farmers to gain access to rainforest areas. For this reason, commercial logging is considered by many to be the biggest single agent of tropical deforestation"

Apart from its direct impact, logging plays a major role in deforestation through the building of roads which are subsequently used by landless farmers to gain access to rainforest areas. These displaced people then clear the forest by slashing and burning to grow enough food to keep them and their families alive, a practice which is called subsistence farming. This problem is so widespread that Robert Repetto of the World Resources Institute ranks commercial logging as the biggest agent of tropical deforestation. This view was supported by the World Wide Fund for Nature's 1996 study, *Bad Harvest?*, which surveyed logging in the world's tropical forests.

Most of the rainforest timber on the international market is exported to rich countries. There, it is sold for hundreds of times the price that is paid to the indigenous people whose forests have been plundered. The timber is used in the construction of doors, window frames, crates, coffins, furniture, plywood sheets, chopsticks, household utensils and other items.

Solutions: For all purposes for which tropical timber is used, other woods or materials could be substituted.

We can stop using tropical timber and urge others to do the same. As long as there is a market for tropical timbers, trees will continue to be cut down. Labelling schemes, aimed at helping consumers to chose environmental friendly timbers.The Rainforest Information Centre provides information to consumers wishing to avoid tropical and other environmentally damaging timbers. Look at the Good Wood Guide on our website, http://rainforestinfo.org.au.

Agriculture - Shifted Cultivators

'Shifted cultivators' is the term used for people who have moved into rainforest areas and established small-scale farming operations. These are the landless peasants who have followed roads into already damaged rainforest areas. The additional damage they are causing is extensive. Shifted cultivators are currently being blamed for 60% of tropical forest loss (Colchester & Lohmann).

The reason these people are referred to as 'shifted' cultivators is that most of them people have been forced off their own land. In Guatemala, rainforest land was cleared for coffee and sugar plantations. The indigenous people had their land stolen by government and corporations. They became 'shifted cultivators', moving into rainforest areas of which they had no previous knowledge in order to sustain themselves and their families (Colchester & Lohmann).

Large-scale agriculture, logging, hydroelectric dams, mining, and industrial development are all responsible for the dispossession of poor farmers. "One of the primary forces pushing landless migrants into the forests is the inequitable distribution of agricultural land" (WRI 1992, Colchester & Lohmann). In Brazil, approximately 42% of cultivated land is owned by a mere 1% of the population. Landless peasants make up half of Brazil's population (WRM).

Once displaced, the 'shifted cultivators' move into forest areas, often with the encouragement of their government. In Brazil, a slogan was developed to help persuade the people to move into the forests. It read "Land without men for men without land" (WRM).

After a time, these farmers encounter the same problems as the cash crop growers. The soil does not remain fertile for long. They are forced to move on, to shift again, going further into the rainforest and destroying more and more of it.

It is evident that the shifted cultivators "have become the *agents* for destruction but not the *cause*" (Westoby 1987: Colchester). Shifted cultivators do not move into pristine areas of undisturbed rainforests.

They follow roads made principally for logging operations. "Shifted cultivators are often used by the timber industry as scapegoats" (Orams and McQuire). Yet logging roads lead to an estimated 90% of the destruction caused by the slash-and-burn farmers (Martin 1991: Colchester).

Solutions: Land reform is essential if this problem is to be addressed. However, according to Colchester and Lohmann, "an enduring shift of power in favour of the peasants" is also needed for such reforms to endure (Colchester &Lohmann).

Agriculture - Cash Crops and Cattle Ranching

Undisturbed and logged rainforest areas are being totally cleared to provide land for food crops, tree plantations or for grazing cattle (Colchester & Lohmann). Much of this produce is exported to rich industrialised countries and in many cases, crops are grown for export while the local populace goes hungry.

Due to the delicate nature of rainforest soil and the destructive nature of present day agricultural practices, the productivity of cash crops grown on rainforest soils declines rapidly after a few years.

Monoculture plantations - those that produce only one species of tree or one type of food - on rainforest soil are examples of non-sustainable agriculture.

They are referred to as cash crops because the main reason for their planting is to make money quickly, with little concern about the environmental damage that they are causing.

Modern machinery, fertilisers and pesticides are used to maximise profits. The land is farmed intensively. In many cases, cattle damage the land to such an extent that it is of no use to cattle ranchers any more, and they move on, destroying more and more rainforest. Not only have the forests been destroyed but the land is exploited, stripped of nutrients and left barren, sustaining no-one.

Solutions:"Reducing the demand for Southern-produced agribusiness crops and alleviating the pressure from externally-financed development projects and assistance is the essential first step" (Colchester and Lohmann).

Fuelwood

The United Nation's Food and Agriculture Organisation estimates that '1.5 billion of the 2 billion people worldwide who rely on fuelwood for cooking and heating are overcutting forests'. This problem is worst in drier regions of the tropics. Solutions will probably involve a return to local peoples' control of the forests they depend on.

Large Dams

In India and South America, hundreds of thousands of hectares of forests have been destroyed by the building of hydro-electric dams. It was the dominant view that new dams had to be built or otherwise these countries would suffer an energy crisis. However, a recent study by the World Bank in Brazil has shown that 'sufficient generating capacity already exists to satisfy the expected rise in demand for power over the medium term, provided that the energy is used more efficiently' (WRM).

The construction of dams not only destroys the forest but often uproots tens of thousands of people, destroying both their land and their culture. The rates of waterborne diseases increase rapidly.

Downstream ecosystems are damaged by dams which trap silt, holding back valuable nutrients. Reduced silt leads to coastal erosion. The sheer weight of water in dams has in Chile, Zimbabwe, and Greece led to earthquakes. The irrigation and industrial projects powered by dams lead to further environmental damage. Irrigation leads to salination of soils and industry leads to pollution.

Solutions: Aid organisations like the World Bank have traditionally favoured spectacular large-scale irrigation and hydro-electric projects. In all cases when such projects are proposed, there has been massive opposition from local people. Reform of the World Bank and other such organisations, and support for campaigns against large-scale dams is needed.

Mining and Industry

Mining and industrial development lead to direct forest loss due to the clearing of land to establish projects. Indigenous people are displaced. Roads are constructed through previously inaccessible land, opening up the rainforest. Severe water, air and land pollution occurs from mining and industry.

Solutions: Local campaigns against mining and industrial development, and the campaigns to reform the large aid agencies which fund such schemes, should be supported.

Colonisation Schemes

Governments and international aid agencies for a time believed that by encouraging colonisation and trans-migration schemes into rainforest areas, they could alleviate some of the poverty felt by the people of the financially poorer countries.

It has since become increasingly obvious that such schemes have failed, hurting the indigenous people and the environment (Colchester & Lohmann). These schemes involve the relocation of millions of people into sparsely populated and forested areas.

In Indonesia, the *Transmigrasi* Program, begun in 1974, is believed to be 'the greatest cause of forest loss in Indonesia', directly causing an average annual loss of 200,000 hectares (Colchester & Lohmann).

The resettled people suffered the same problems as 'shifted cultivators'. The soil is not fertile enough to be able to sustain them for very long.

Even after such projects have officially ended, the flow of 'shifted cultivators' continues as the area remains opened up. "The World Bank estimates that for every colonist resettled under the official transmigration project, two or more unofficially move into the forest due to the drawing effect of the program" (Colchester & Lohmann).

Tourism

The creation of national parks has undoubtedly helped to protect rainforests. Yet, as national parks are open to the public, tourism is damaging some of these areas. Often, national parks are advertised to tourists before adequate management plans have been developed and implemented. Inadequate funding is allocated for preservation of forests by government departments. Governments see tourism as an easy way to make money, and therefore tourism is encouraged whilst strict management strategies are given far less government support.

Ecotourism, or environmentally friendly tourism, should educate the tourists to be environmentally aware. It should also be of low impact to its environment. Unfortunately, many companies and resorts who advertise themselves as eco-tourist establishments are in fact exploiting the environment for profit.

In Cape Tribulation, Australia the rainforest is being threatened by excessive tourism. Clearing for roads and pollution of waterways are two of the major problems in this area. The Wet Tropics Management Authority which oversees the surrounding World Heritage Area is promoting tourism to the area before any management plans have been formulated, before any effective waste management strategy has been devised and before any ecofriendly power alternatives have been fully explored. Solutions: The rights of indigenous forest dwellers and others who depend on intact forests must be upheld. In instances where there are campaigns opposing specific tourist developments, they should be supported. Genuine ecotourism should be preferred to other tourist enterprises.

UNDERLYING CAUSES

More Than Just Poverty and Overpopulation

Poverty and overpopulation are believed to be the main causes of forest loss, according to the international agencies such as the FAO and intergovernmental bodies. They believe they can solve the problem by encouraging development and trying to reduce population growth. However, the World Rainforest Movement and many other non-governmental organisations hold unrestrained development and the excessive consumption habits of rich industrialised countries directly responsible for most forest loss.

Development and Overconsumption: the Basic Causes

The World Rainforest Movement's Emergency Call to Action for the Forests and Their Peoples asserts that "deforestation is the inevitable result of the current social and economic policies being carried out in the name development". It is the push for development which gives rise to commercial logging, cash crops, cattle ranching, large dams, colonisation schemes, the dispossession of peasants and indigenous people and the promotion of tourism.

Harrison Ngau, an indigenous tribesman from Sarawak, Malaysia and winner of the Goldman Environment Award in 1990, has this to say about why tropical forests are being destroyed:

The roots of the problem of deforestation and waste of resources are located in the industrialised countries, where most of our resources, such as tropical timber end up. The rich nations with one quarter of the world's population consume four fifth of the world's resources. It is the throw away culture of the industrialised countries, now advertised in and forced on to the Third World countries that is leading to the throwing away of the world. Such so-called progress leads to destruction and despair![World Rainforest Movement]

Colonialism

Tropical rainforests are found mainly in the Third World countries, Australia and Hawaii being the only exceptions. All of these countries have indigenous populations who had their own system of land management and/or ownership in place for thousand of years before the intervention of colonists from rich industrialised nations. The colonial powers (Britain,

France, Spain and Portugal), whilst exploiting the resources of many of these countries, attempted to destroy indigenous peoples' rights to remain on their land. Colonialism turned previously self-sufficient economies into zones of agriculture export production (Colchester and Lohmann). This process continues today and the situation is worsening.

Exploitation by Industrialised Countries

Wealthy countries have been consuming so much of their own resources that they are no longer sustaining their growing populations and increasingly, they are turning to the resources of the financially poorer countries. "Twenty per cent of the world's population is using 80 % of the world's resources" (Orams & McQuire).

Currently, although many indigenous people are claiming their culture and rights, they face stubborn opposition, as the governments in their own countries have often 'adopted the same growth-syndrome as their Western neighbours, with the emphasis on maximising exports, revenues and exploiting resources for short-term gain. Corruption in government, the military and economic powers is well known' (Orams & McQuire). The problem is made worse by the low price for most Third World exports on the international market. The United States has been accused of manipulating prices for agricultural commodities for its own benefit at the expense of tropical countries (WRR).

The Debt Burden

The governments of the financially poorer countries feel they need to make money in order to repay their huge international debts. In the 1970's and 80's, they borrowed vast sums of money from development agencies in industrialised countries in order to improve their own economies. Most are still battling to make repayments due to escalating interest rates (Orams & McQuire). Since 1987, the flow of debt repayments from Third World countries to rich countries has exceeded the flow of aid money going to Third World countries (RIC). Poor countries feel compelled to exploit their natural resources, including their forests, partly to earn foreign exchange for servicing their debts. Non-government organisations in Third World countries have for many years been pointing out that there is no chance of stopping impoverishment and destruction of nature without a solution to the debt crisis.

In some countries in South-East Asia, the construction of roads for logging operations was funded by Japanese aid. Later, the forests were exploited by Japanese timber companies. The timber companies made the profits and the South-East Asian countries were left owing Japan money for the construction of the roads (Orams & Maquire).

The Role of Poverty and Overpopulation

Poverty, while undeniably responsible for much of the damage to rainforests, has to a large extent been brought about by the greed of the rich industrialised nations and the Third World elites who seek to emulate them. Development, which is often seen as the solution to world poverty, seldom helps those whose need is greatest. It is often the cause rather than the cure for poverty.

The claim that overpopulation is the cause of deforestation is used by many governments and aid agencies as an excuse for inaction. In tropical countries, pressure from human settlement comes about more from inequitable land distribution that from population pressure. In general, most of the land is owned by a small but powerful elite which displaces poor farmers into rainforest areas. So long as these elites maintain their grip on power, lasting land reform will be difficult to achieve.

Overpopulation is not a problem exclusive to Third World countries. An individual in an industrialised country is likely to consume in the order of sixty times as much of the world's resources as a person in a poor country. The growing populations in rich industrialised nations are therefore responsible for much of the exploitation of the earth, and there is a clear link between the overconsumption in rich countries and deforestation in the tropics.

RAIN FOREST CHARACTERISTICS

Tropical rain forests belong to a broader forest category called tropical moist forests, of which there are many different types. Forest scientists distinguish rain forests from other types by considering factors such as temperature, rainfall, length of dry season, and altitude.

Temperature

Rain forests are typically hot and steamy—the average annual temperature is 25° C (77° F). Temperature near the equator varies little over the course of a year, so rain forest temperatures are about the same year round—the average minimum monthly temperature in a rain forest is a balmy 18° C (64° F). Rain forests are not found where the temperature drops close to 0° C (32° F) because their resident plants and animals are not well adapted to withstand frost. Temperature in a rain forest depends not only on distance from the equator but also on altitude.

As elevation increases, nighttime temperatures go down significantly. This daily temperature variation affects forest ecology, and true rain forests typically are not found above 1,000 m (3,280 ft).

Rainfall

Rain forests can average as little as 1.8 m (6 ft), or as much as 9.0 m (30 ft), of rainfall a year. What distinguishes a true rain forest is the distribution of precipitation throughout the year—there are no dry seasons. Every month, typically more than 100 mm (4 in) of rain falls. If a rain forest does have dry periods, they are usually short and unpredictable.

In many climates, rainfall evaporation is carried away to fall as rain in distant places, but in the rain forests, nearly 50 percent of the precipitation comes from local evaporation. The warm, humid air surrounding a rain forest forms a microenvironment that permits little water to escape. Much of the rain that falls on the rain forest is intercepted by the trees in the canopy. Some of it rolls off the leaves and down the trunks to the forest below, but a high percentage evaporates and hangs as tiny droplets of water in the humid atmosphere. Gentle and continuous winds lift the tiny droplets higher in the atmosphere, where they cool and form clouds. When enough of these drops enter the atmosphere and cool, they condense and fall as rain, beginning the cycle again.

Soil

Despite their incredible lushness and high diversity, one of the peculiarities of rain forests is that the soil is poor in nutrients that can be absorbed by plant roots. The nutritious minerals have been washed out of the soils by heavy rainfall and high temperatures over thousands of years. To compensate for the nutrient-poor rain forest soils, most tropical trees absorb the nutrients they can find and hold them in their living tissue.

In contrast, the rich soils of temperate forests are better able to retain nutrients, enabling a temperate forest tree to absorb small amounts of minerals as the tree needs them. When tropical trees die, nutrients are released into the soils by decomposition. Rather than remaining in the soil reservoir as they would in a temperate forest, in the rain forest, the nutrients are rapidly absorbed again by other living organisms.

FOREST STRUCTURE

Rain forest structure is distinct from most other forest types because of its many layers of vegetation, referred to as *strata*. The three strata of a rain forest are the understory, the midstory, and the overstory. The lowest stratum is the understory, composed of palms, herbaceous plants (such as wild ginger), and tree seedlings and saplings. Just 2 percent of the sunlight penetrates the many layers of leaves and branches above, so understory plant species have developed special traits to cope with low light levels. Many have deep red colouring on the underside of their leaves to capture some of the scarce light that does manage to reach the forest understory. This red colouring enables understory plants to absorb light of different wavelengths than the lush, greenfoliaged canopy plants do. Above the forest floor but below the canopy are one or more midstory strata, made up of woody plants, such as large shrubs and midsized trees.

The overstory is the canopy, in which the tree crowns form a continuous layer that captures the bulk of the rainwater and sunlight hitting the forest. The height of the canopy varies from region to region and forest to forest, ranging from 20 to 50 m (65 to 165 ft).

The lush, green canopy is teeming with life, and forest researchers have developed ingenious methods for accessing this mysterious ecosystem. Researchers use hot air balloons, cables, catwalks, towers, sophisticated tree-climbing gear, and even robots to study the millions of plants and animals that make their home high up in the forest canopy. Canopy researchers also use huge cranes that are dropped into the heart of the forest by helicopters. Suspended from the crane's long, movable arm is a large gondola that functions as a mobile treetop laboratory. Moving from tree to tree, forest researchers collect specimens, conduct experiments, and observe life in the canopy frontier.

The highest stratum of the rain forest is made up of the emergent trees, those individuals that stick up above the forest canopy. Emergents, which do not form a continuous layer, are usually the giants of the forest, reaching heights of 35 to 70 m (115 to 230 ft) or more, and trunk girths of over 2 m (6.6 ft) in diameter. Less than one percent of the trees in the forest reside in the canopy and emergent layers. However, these trees tend to be so large that they collectively account for the vast majority of the woody mass, or biomass, of the forest.

The nicely ordered strata of the rain forest, including the continuous layer of the canopy, are regularly disturbed by naturally occurring events, such as falling trees. Trees in a rain forest canopy are often interconnected by vines, and a falling tree may pull as well as push other trees down with it, producing a domino effect of falling trees. The resulting opening in the forest canopy enables light to pour onto the forest floor. New plants and animals then move into the area and begin to grow.

Other natural disturbances create even larger openings in the forest canopies. Along the hurricane belt in the Caribbean and the typhoon belt along the western Pacific, some forests are substantially altered when high winds and storms blow down hundreds of trees every few decades. On a smaller scale, large mammals, such as elephants, regularly raze rain forest vegetation in the Congo River Basin in Africa. Scientists have found that these natural disturbances and the subsequent forest regeneration are a vital process that leads to healthy and diverse forests.

PLANT AND ANIMAL LIFE

Rain forest ecosystems contain more plant and animal species than virtually any other habitat in the world. Although their range has contracted and expanded with climate changes over the last several million years, in general, rain forests are some of the oldest ecosystems on Earth. As a result of this continuity, rain forests boast millions of different species, many of which are *endemic*, or unique, to rain forest habitats.

Plants

Although they contain numerous species, rain forests are remarkably uniform in their general appearance. Most trees have tall, slender trunks that do not branch until near the crown. Many, such as kapok trees, are supported by thick buttresses that can stretch out 10 m (33 ft) or more. These buttresses provide needed support for rain forest trees, top heavy because nutrient-poor rain forest soils lead to fragile, shallow root systems. Rain forest tree bark tends to be thin and smooth. Notable exceptions are palms, which are common in some rain forests and virtually absent in other types of forests.

Rain forest plants have many unique physical characteristics that exploit the particular habitat, or niche, that a species occupies. Understory and midstory plants, such as relatives of the banana tree, tend to have particularly large leaves to capture as much light as possible—what little light that is not intercepted by the canopy above. These large leaves do not dry out as they would high in the canopy, where the intense sunlight creates a drier environment. These traits can change, however, when a plant's environment changes. Canopy trees may change their shape over the course of their life, depending on the environment around them. Leaves often get smaller as trees grow larger. In some cases, leaves of juvenile plants may be almost 10 times larger than adults of the same canopy tree species.

On the nutrient-poor forest floor, many rain forest plants, such as human-sized members of the palm family called *Astrocaryum sciophilum,* collect falling debris from other plants in their cup-shaped leaves to create their own little compost pile. Rain forests also feature insectivorous plants, which derive some of their nutrients by trapping animals, particularly insects, in their leaves.

Among the insectivorous plants are the pitcher plants native to tropical Asia. Insects land upon the pitcher plant's tubular-shaped leaves, then slide into a cavity, or *pitcher,* at the plant's centre that is full of digestive juices. Here, the insect quickly dissolves and the nutrients it contains become available for the plant's use.

Rain forests are also home to the largest flower in the world, the giant rafflesia flower, commonly known as the corpse lily. Weighing up to 7 kg (15 lb) with petals spanning almost 1 m (1 yd), the giant rafflesia flower is best known for its putrid smell. The plant emits the stench of rotting meat to attract certain flies, who serve as its pollinators.

Rain forests are often teeming with climbing plants, such as rattan palms. These thick, woody climbers—25 cm (10 in) in diameter—are frequently found connecting the trees, ascending up into the treetops and looping back down. By climbing trees, these *lianas,* or vines, expose their leaves and flowers to sunlight, birds, and insects without expending the energy needed to build their own supporting tissues. Epiphytes, including mosses, bromeliads, and orchids, grow on tree trunks or nestled in the crook of a tree. Lacking permanent roots in the soil, epiphytes must obtain their nutrients from other living plants or by trapping water and organic matter as they fall to the forest floor.

Bromeliads can store up to 38 liters (10 gallons) of water in the reservoirs formed by their overlapping leaves. Most live harmoniously with their host trees, though some are less benign. Strangler figs, which begin their lives as epiphytes, germinate in the canopy trees and send roots down to the soil. As they grow, these parasites enwrap their hosts until they literally strangle them to death, at which point the figs become free-standing trees.

Animals

Almost 90 percent of the rain forest animal species are insects, and of these, most are beetles. A single rain forest tree can host more than 150 species of beetles. Living high in the forest canopy, most of these beetles and other insect species have eluded scientists until recently, when technology has improved access to the upper stratum. To this day, scientists are unsure how many animal species exist in the world, largely because they have identified just a small fraction of the millions—some estimate as many as 30 million—of insects that live in the rain forest.

Among the most fascinating rain forest insects yet encountered are leafcutter ants, remarkable because they actually cultivate their own food. These ants cut the leaves of particular plants and carry them back to their underground nests, where they fertilize them with saliva. This careful tending causes growth of a particular fungus, which the ants harvest and rely on as their sole source of food.

In 15 sq km (6 sq mi) of rain forest, as many as 100 different mammal species may be found.

These animals occupy every available niche, from burrows in the forest floor to the branches of emergent trees. Most rain forest mammals are nocturnal (active during the night) or crepuscular (active during the twilight of dusk or dawn), and they spend the heat of the day sleeping. In fact, almost half the mammals of the rain forest are bats, flying mammals known for their nighttime activity.

Some rain forest mammals, including gorillas, elephants, tapirs, agoutis, and wild pigs, are ground-dwellers, but most, like their insect counterparts, live high in the treetops in the forest canopy. Canopy-dwellers have evolved an array of fascinating traits to survive in the branches of trees. For instance, some Asian rain forests are especially known for animals with the ability to glide. Borneo alone has more than 30 species of mammals, reptiles, and amphibians that can glide from one tree to another. Most rain forest monkeys in Central and South America use their specialized tails, which are capable of grasping branches, like a fifth limb while climbing, feeding, and even playing high above the forest floor.

The three-toed sloth spends most of its life defenceless, hanging upside down from tree branches. To avoid predators, its movements are so slow that they are virtually undetectable to would-be predators, even vigilant jaguars. The sloth has also developed a relationship with a rain forest plant that makes it even more elusive to its predators: Although it has brown fur, the sloth blends in with the surrounding green canopy because a particular species of green alga lives in its fur.

Environment, all of the external factors affecting an organism. These factors may be other living organisms (biotic factors) or nonliving variables (abiotic factors), such as temperature, rainfall, day length, wind, and ocean currents. The interactions of organisms with biotic and abiotic factors form an ecosystem. Even minute changes in any one factor in an ecosystem can influence whether or not a particular plant or animal species will be successful in its environment.

Organisms and their environment constantly interact, and both are changed by this interaction. Like all other living creatures, humans have clearly changed their environment, but they have done so generally on a grander scale than have all other species.

Some of these human-induced changes—such as the destruction of the world's tropical rain forests to create farms or grazing land for cattle—have led to altered climate patterns. In turn, altered climate patterns have changed the way animals and plants are distributed in different ecosystems.

Scientists study the long-term consequences of human actions on the environment, while environmentalists—

professionals in various fields, as well as concerned citizens—advocate ways to lessen the impact of human activity on the natural world.

UNDERSTANDING THE ENVIRONMENT

The science of ecology attempts to explain why plants and animals live where they do and why their populations are the sizes they are. Understanding the distribution and population size of organisms helps scientists evaluate the health of the environment. Population size and distribution may also be affected, either directly or indirectly, by the way species in an ecosystem interact with one another.

In an experiment performed in the late 1960s in the rocky tidal zone along the Pacific Coast of the United States, American ecologist Robert Paine studied an area that contained 15 species of invertebrates, including starfish, mussels, limpets, barnacles, and chitons.

Paine found that in this ecosystem one species of starfish preyed heavily on a species of mussel, preventing that mussel population from multiplying and monopolizing space in the tidal zone. When Paine removed the starfish from the area, he found that the mussel population quickly increased in size, crowding out most other organisms from rock surfaces. The number of invertebrate species in the ecosystem soon dropped to eight species.

Paine concluded that the loss of just one species, the starfish, indirectly led to the loss of an additional six species and a transformation of the ecosystem.Typically, the species that coexist in ecosystems have evolved together for many generations.

These populations have established balanced interactions with each other that enable all populations in the area to remain relatively stable.

Occasionally, however, natural or human-made disruptions occur that have unforeseen consequences to populations in an ecosystem.

17th-century sailors routinely introduced goats to isolated oceanic islands, intending for the goats to roam freely and serve as a source of meat when the sailors returned to the islands during future voyages.

As nonnative species free from all natural predators, the goats thrived and, in the process, overgrazed many of the islands. With a change in plant composition, many of the native animal species on the islands were driven to extinction. A simple action, the introduction of goats to an island, yielded many changes in the island ecosystem, demonstrating that all members of a community are closely interconnected.

EROSION

Erosion, removal of rock and soil material by natural processes, principally running water, glaciers, waves, and wind. Erosion transports rocky material after the process of weathering has broken bedrock down into smaller, moveable pieces.Through erosion the surface of the earth is constantly being sculptured into new forms.

The shapes of continents are continuously changing, as waves and tides cut into old land while silt from rivers builds up new land. As rivulets, streams, and rivers cut their channels deeper, gullies become ravines and ravines become valleys. The Grand Canyon, more than 1500 m (more than 5000 ft) deep, was produced by erosion probably within the past 5 million years.

The overall effect of the wearing down of mountains and plateaus is to level the land; the tendency is toward the reduction of all land surfaces to sea level. In each 7000- to 9000-year period the entire area drained by the Mississippi River loses an average of 30 cm (12 in) of altitude. Opposing this tendency are volcanic eruptions and movements of the crust of the earth that raise mountains, plateaus, and new islands.

WEATHERING

Weathering initiates the erosion of rock, causing alterations in the surface layers. In dry climates, the top layer of a rock may expand from the heat of the sun and crack off from the lower layers. If the rock consists of several minerals, the minerals may expand at different rates and break up the rock. In cold climates, frost breaks up rocks because rainwater, which seeps into cracks and pores in the rock, expands when it freezes.

Rain in damp climates acts chemically as well as mechanically in the weathering of rocks. As the rain passes through the atmosphere it absorbs carbon dioxide, forming carbonic acid, which dissolves some minerals and decomposes others. Feldspar, a common family of minerals in granite, is changed into clays, and certain minerals in basalt combine with oxygen and water to form iron oxides, such as limonite. Plants play a role in weathering as roots can split rocks and extract soluble nutrients.

Water Erosion

Water plays an important role in erosion by carrying away material that has been weathered and broken down. When an area receives more water (in the form of rain, melting snow, or ice) than the ground can absorb, the excess water flows to the lowest level, carrying loose material with it. Gentle slopes are subject to *sheet* and *rill* erosion, in which the runoff removes a thin layer of topsoil without leaving visible traces on the eroded surface. This erosion may be balanced by the formation of new soil.

Often, however, especially in arid areas having little vegetation, the runoff leaves a pattern of gullies formed by rivulets. Water can even erode solid rock, especially along streambeds where the stones that are carried with the current scour and abrade. Every year rivers deposit about 3.5 million tons of eroded material into the oceans.

Glacial Erosion

Glaciers are important agents of erosion. Although a glacier moves slowly, it gradually removes all the loose material from the surface over which it travels, leaving bare rock surfaces when the ice melts. Besides removing loose material, glaciers actively erode the solid rock over which they travel. Rock fragments that become embedded in the bottom and sides of the moving ice mass act as an abrasive, grinding and scouring the bedrock which forms the walls and floors of mountain valleys.

Coastal Erosion

Coastal erosion of rocky cliffs and sandy beaches results from the action of ocean waves and currents. This is especially severe during storms. In many parts of the world the loss of land due to coastal erosion represents a serious problem. The action of waves, however, does not extend to a great depth, and the sea tends to cut a flat platform, characteristic of marine erosion, into coastal rocks.

Wind Erosion

Wind is another active agent of erosion, especially in arid climates with little vegetation. Wind blowing across bare land lifts particles of sand and silt but leaves behind larger pebbles and cobbles. Eventually, a surface layer of closely packed stones, called a desert pavement, is formed as the sand and silt is removed.

The removal of large quantities of loose material is called deflation. Deflation lowers the landscape slowly, usually less than a meter (3 ft) in a thousand years. However, deflation can occur more rapidly, as it did during the 1930s in the southern parts of the Great Plains of the United States. Winds may sometimes deposit sand in large piles, known as sand dunes.

Human Impacts on Erosion

Without human activities, losses of soil through erosion would in most areas probably be balanced by the formation of new soil. On virgin land a mantle of vegetation protects the soil. When rain falls on a surface of grass or on the leaves of trees, some of the moisture evaporates before it can reach the ground. Trees and grass serve as windbreaks, and a network of roots helps to hold the soil in place against the action of both rain and wind.

Agriculture and lumbering, as well as housing, industrial development, and highway construction, however, partially or wholly destroy the protective canopy of vegetation and greatly speed up erosion of certain kinds of soils. Erosion is less severe with crops such as wheat, which cover the ground evenly, than with crops such as corn and tobacco, grown in rows.

Overgrazing, which in time can change grassland to desert, and careless cultivation have had disastrous effects in certain parts of the United States. Some historians believe that soil erosion has been a determinant in the complex of causes underlying various population shifts and the fall of certain civilizations. Ruins of towns and cities have been found in arid regions, such as the deserts of Mesopotamia, indicating that agriculture was once widespread in the surrounding territory.

Global Warming or Climate Change, measurable increases in the average temperature of Earth's atmosphere, oceans, and landmasses. Scientists believe Earth is currently facing a period of rapid warming brought on by rising levels of heat-trapping gases, known as greenhouse gases, in the atmosphere.

Greenhouse gases retain the radiant energy (heat) provided to Earth by the Sun in a process known as the greenhouse effect. Greenhouse gases occur naturally, and without them the planet would be too cold to sustain life as we know it. Since the beginning of the Industrial Revolution in the mid-1700s, however, human activities have added more and more of these gases into the atmosphere. Levels of carbon dioxide, a powerful greenhouse gas, have risen by 35 percent since 1750, largely from the burning of fossil fuels such as coal, oil, and natural gas. With more greenhouse gases in the mix, the atmosphere acts like a thickening blanket and traps more heat.

GLOBAL WARMING IN THE PAST

Earth has warmed and cooled many times since its formation about 4.6 billion years ago. Global climate changes were due to many factors, including massive volcanic eruptions, which increased carbon dioxide in the atmosphere; changes in the intensity of energy emitted by the Sun; and variations in Earth's position relative to the Sun, both in its orbit and in the inclination of its spin axis.

Variations in Earth's position, known as Milankovitch cycles, combine to produce cyclical changes in the global climate. These cycles are believed to be responsible for the repeated advance and retreat of glaciers and ice sheets during the Pleistocene Epoch (1.8 million to 11,500 years before present), when Earth went through fairly regular cycles of colder "glacial" periods (also known as ice ages) and warmer "interglacial" periods. Glacial periods occurred at roughly 100,000-year intervals.

An interglacial period began about 10,000 years ago, when the last ice age came to an end. Prior to that ice age, an interglacial period occurred about 125,000 years ago. During interglacial periods, greenhouse gases such as carbon dioxide and methane naturally increase in the atmosphere from increased plant and animal life. But since 1750 greenhouse gases have increased dramatically to levels not seen in hundreds of thousands of years, due to the rapid growth of the human population combined with developments in technology and agriculture. Human activities now are a powerful factor influencing Earth's dynamic climate.

The ice of the polar regions furnishes clues to the makeup of Earth's ancient atmosphere. Ice cores that scientists have bored from the ice sheets of Greenland and Antarctica provide natural records of both temperature and atmospheric greenhouse gases going back hundreds of thousands of years. Layers in these ice cores created by seasonal snowfall patterns allow scientists to determine the age of the ice in each core.

By measuring tiny air bubbles trapped in the ice and properties of the ice itself, scientists can estimate the temperature and amount of greenhouse gases in Earth's past atmosphere at the time each layer formed. Based on this data, scientists know that greenhouse gases have now risen to levels higher than at any time in the last 650,000 years.Greenhouse gases are rising, and temperatures are following.Before the late 1800s, the average surface temperature of Earth was almost 15°C (59°F).

Over the past 100 years, the average surface temperature has risen by about 0.7 Celsius degrees (1.3 Fahrenheit degrees), with most of the increase occurring since the 1970s. Scientists have linked even this amount of warming to numerous changes taking place around the world, including melting mountain glaciers and polar ice, rising sea level, more intense and longer droughts, more intense storms, more frequent heat waves, and changes in the life cycles of many plants and animals. Warming has been most dramatic in the Arctic, where temperatures have risen almost twice as much as the global average.

GLOBAL WARMING IN THE FUTURE

Scientists project global warming to continue at a rate that is unprecedented in hundreds of thousands or even millions of years of Earth's history. They predict considerably more warming in the 21st century, depending on the level of future greenhouse gas emissions.

For a scenario (possible situation) assuming higher emissions—in which emissions continue to increase significantly during the century—scientists project further warming of 2.4 to 6.4 Celsius degrees (4.3 to 11.5 Fahrenheit degrees) by the year 2100. For a scenario assuming lower emissions—in which emissions grow slowly, peak around the year 2050, and then fall—scientists project further warming of 1.1 to 2.9 Celsius degrees (1.9 to 5.2 Fahrenheit degrees) by the year 2100.

Melting polar ice and glaciers, as well as warming of the oceans, expands ocean volume and raises sea level, which will eventually flood some coastal regions and even entire islands. Patterns of rainfall are expected to change, with higher latitudes (closer to the poles) projected to receive more rainfall, and subtropical areas (such as the Mediterranean and southern Africa) projected to receive considerably less. Changes in temperature and precipitation patterns may damage food crops, disrupting food production in some parts of the world. Plant and animal species will shift their ranges toward the poles or to higher elevations seeking cooler temperatures, and species that cannot do so may become extinct. Increasing levels of carbon dioxide in the atmosphere also leads to increased ocean acidity, damaging ocean ecosystems.

Human beings face global warming with a huge population at risk. The potential consequences are so great that many of the world's leading scientists—and increasingly, politicians, business leaders, and other citizens—are calling for international cooperation and immediate action to counteract the problem.

THE GREENHOUSE EFFECT

The energy that lights and warms Earth comes from the Sun. Short-wave radiation from the Sun, including visible light, penetrates the atmosphere and is absorbed by the surface, warming Earth. Earth's surface, in turn, releases some of this heat as long-wave infrared radiation.

Much of this long-wave infrared radiation makes it back out to space, but a portion remains trapped in Earth's atmosphere, held in by certain atmospheric gases, including water vapour, carbon dioxide, and methane. Absorbing and reflecting heat radiated by Earth, these gases act somewhat like the glass in a greenhouse, and are thus known as greenhouse gases.

Only greenhouse gases, which make up less than 1 percent of the atmosphere, offer the Earth any insulation. All life on Earth relies on the greenhouse effect—without it, the average surface temperature of the planet would be about -18°C (0°F) and ice would cover Earth from pole to pole.

TYPES OF GREENHOUSE GASES

Greenhouse gases occur naturally in the environment and also result from human activities. By far the most abundant greenhouse gas is water vapour, which reaches the atmosphere through evaporation from oceans, lakes, and rivers. The amount of water vapour in the atmosphere is not directly affected by human activities.

Carbon dioxide, methane, nitrous oxide, and ozone all occur naturally in the environment, but they are being produced at record levels by human activities. Other greenhouse gases do not occur naturally at all and are produced only through industrial processes. Human activities also produce airborne particles called aerosols, which offset some of the warming influence of increasing greenhouse gases.

CARBON DIOXIDE

Carbon dioxide is the second most abundant greenhouse gas, after water vapour. Carbon dioxide constantly circulates in the environment through a variety of natural processes known as the carbon cycle. It is released into the atmosphere from natural processes such as eruptions of volcanoes; the respiration of animals, which breathe in oxygen and exhale carbon dioxide; and the burning or decay of plants and other organic matter. Carbon dioxide leaves the atmosphere when it is absorbed into water, especially the oceans, and by plants, especially trees. Through a process called photosynthesis, plants use the energy of light to convert carbon dioxide and water into simple sugars, which they use as food. In the process, plants store carbon in new tissue and release oxygen as a byproduct.

Humans are significantly increasing the amount of carbon dioxide released to the atmosphere through the burning of fossil fuels (such as coal, oil, and natural gas), solid wastes, and wood and wood products to heat buildings, drive vehicles, and generate electricity. At the same time, the number of trees available to absorb carbon dioxide through photosynthesis has been greatly reduced by deforestation, the widespread cutting of trees for lumber or to clear land for agriculture.

Human activities are causing carbon dioxide to be released to the atmosphere much faster than Earth's natural processes can remove it. In addition, carbon dioxide can remain in the atmosphere a century or more before nature can dispose of it. Before the Industrial Revolution began in the mid-1700s, there were about 280 molecules of carbon dioxide per million molecules of air (abbreviated as parts per million, or ppm). Concentrations of carbon dioxide have risen since then as industrial production and fossil fuel-based transportation and electricity generation have spread around the world, accelerating in the last 50 years. In 2007 the Intergovernmental Panel on Climate Change (IPCC), a major scientific organization, reported that levels of carbon dioxide had risen to a record high of 379 ppm and are increasing an average of 1.9 ppm per year.

To stabilize atmospheric concentrations of carbon dioxide, global emissions would need to be cut significantly—on the order of 70 to 80 percent. If efforts are not made to reduce greenhouse gas emissions, carbon dioxide is projected to reach concentrations more than double or even triple the level prior to the Industrial Revolution by 2100.

In a higher-emissions scenario carbon dioxide is projected to reach 970 ppm by 2100, more than tripling preindustrial concentrations. In a lower-emissions scenario, carbon dioxide is projected to reach 540 ppm by 2100, still almost doubling preindustrial concentrations. (For a description of these two emissions scenarios.

METHANE

Methane is emitted into the atmosphere during the mining of coal and the production and transport of natural gas and oil. Methane also comes from rotting organic matter in landfills, rice paddies, and wetlands, as well as from certain animals, especially cows, as a byproduct of digestion. Live plants also emit small amounts of methane.

Scientists are increasingly concerned about the release of methane and carbon dioxide from melting permafrost, areas of frozen ground in the tundra (Arctic plains) of Alaska, Siberia, and other subpolar regions. Temperatures in the top layer of permafrost have increased, leading to a decrease in the area of seasonally frozen ground. Methane released from these areas as they melt would contribute to further warming and further melting, in what scientists call a feedback process.

Since the beginning of the Industrial Revolution, the amount of methane in the atmosphere has more than doubled. Methane traps nearly 30 times more heat than the same amount of carbon dioxide. Compared to carbon dioxide, methane appears in lower concentrations in the atmosphere and remains in the atmosphere for a shorter time. In total, methane contributes about a third as much as carbon dioxide to global warming.

NITROUS OXIDE

Nitrous oxide is a potent greenhouse gas that is released primarily by plowing farm soils and burning fossil fuels. Nitrous oxide traps about 300 times more heat than does the same amount of carbon dioxide. The concentration of nitrous oxide in the atmosphere has increased 18 percent over preindustrial levels. Nitrous oxide contributes about a tenth as much as carbon dioxide to global warming.

OZONE

Ozone is both a natural and human-made greenhouse gas. Ozone in the upper atmosphere is known as the ozone layer and shields life on Earth from the Sun's harmful ultraviolet radiation. This ozone is formed by the action of ultraviolet light from the Sun on molecules of ordinary oxygen. Some chemical compounds are known to destroy ozone molecules in the upper atmosphere. This can break down, or deplete, the ozone layer. Depletion of the ozone layer actually causes a slight cooling, offsetting a small part of the warming from greenhouse gases.

However, ozone in the lower atmosphere is a component of smog, a severe type of air pollution. Nitrogen oxides and volatile organic gases emitted by automobiles and industrial sources combine to form the ozone in smog. This ozone is a poison that damages vegetation, kills trees, irritates lung tissues, and attacks rubber. It is also a greenhouse gas that contributes about a fourth as much as carbon dioxide to global warming. Unlike the greenhouse gases which are well-mixed throughout the atmosphere, ozone in the lower atmosphere tends to be limited to industrialized regions.

Index

A

B

C

D

E

F

G

Q

R

S

T